POLYMER DYNAMICS AND RELAXATION

Polymers exhibit a range of physical characteristics, from rubber-like elasticity to the glassy state. These particular properties are controlled at the molecular level by the mobility of the structural constituents. Remarkable changes in mobility can be witnessed with temperature, over narrow, well defined regions, termed relaxation processes. This is an important, unique phenomena controlling polymer transition behavior and is described here at an introductory level. The important types of relaxation processes from amorphous to crystalline polymers and polymeric miscible blends are covered, in conjunction with the broad spectrum of experimental methods used to study them. In-depth discussion of molecular level interpretation, including recent advances in atomistic level computer simulations and applications to molecular mechanism elucidation are discussed. The result is a self-contained, up-to-date approach to polymeric interpretation suitable for researchers and graduate students in materials science, physics and chemistry interested in the relaxation processes of polymeric systems.

RICHARD H. BOYD is a distinguished Professor Emeritus of Materials Science and Engineering and of Chemical Engineering at the University of Utah. He was awarded the Polymer Physics Prize from the American Physical Society and is the author of over 150 technical papers and book chapters.

GRANT D. SMITH is a Professor at the University of Utah. He is an NSF Career awardee and Humboldt Fellow. He is also the author of co-author of over 170 papers.

POLYMER DYNAMICS AND RELAXATION

RICHARD H. BOYD

*Distinguished Professor Emeritus of Materials Science and
Engineering and of Chemical Engineering, University of Utah*

GRANT D. SMITH

Professor of Materials Science and Engineering, University of Utah

CAMBRIDGE
UNIVERSITY PRESS

CAMBRIDGE UNIVERSITY PRESS
Cambridge, New York, Melbourne, Madrid, Cape Town, Singapore,
São Paulo, Delhi, Dubai, Tokyo, Mexico City

Cambridge University Press
The Edinburgh Building, Cambridge CB2 8RU, UK

Published in the United States of America by Cambridge University Press, New York

www.cambridge.org
Information on this title: www.cambridge.org/9780521152914

First published 2007
First paperback printing 2010

A catalogue record for this publication is available from the British Library

ISBN 978-0-521-81419-5 Hardback
ISBN 978-0-521-15291-4 Paperback

Contents

Preface		*page* ix
Part I	**Methodology**	1
1	Mechanical relaxation	3
	1.1 Regimes of behavior	3
	1.2 Superposition principle	5
	1.3 Relaxation modulus	5
	1.4 Simple stress relaxation	6
	1.5 Dynamic modulus	7
	1.6 Interconversion of stress relaxation and dynamic modulus	9
	1.7 Representation of the relaxation function: single relaxation time (SRT)	11
	1.8 Relaxations in polymeric materials tend to be "broad"	13
	1.9 Distribution of relaxation times	14
	1.10 Relaxation spectrum from $E_R(t)$	15
	1.11 Creep compliance	18
	1.12 Dynamic compliance	19
	1.13 Representation of the retardation function	21
	1.14 Summary of the data transformations illustrated	22
	Appendix A1 A brief summary of elasticity	23
	References	26
2	Dielectric relaxation	27
	2.1 Dielectric permittivity	27
	2.2 Measurement of dielectric permittivity	30
	2.3 Time dependence of polarization: reorientation of permanent dipoles	31
	2.4 Polarization and permittivity in time dependent electric fields	33
	2.5 Empirical representations of the dielectric permittivity	35
	References	43

3   NMR spectroscopy   44
    3.1   NMR basics   45
    3.2   The pulsed NMR method   47
    3.3   NMR relaxation measurements   49
    3.4   NMR exchange spectroscopy   54
          References   56
4   Dynamic neutron scattering   57
    4.1   Neutron scattering basics   57
    4.2   Time-of-flight (TOF) and backscattering QENS   63
    4.3   Neutron spin echo (NSE) spectroscopy   66
          References   69
5   Molecular dynamics (MD) simulations of amorphous polymers   70
    5.1   A brief history of atomistic MD simulations
          of amorphous polymers   70
    5.2   The mechanics of MD simulations   71
    5.3   Studying relaxation processes using atomistic MD simulations   75
    5.4   Classical atomistic force fields   76
          References   79
Part II   Amorphous polymers   81
6   The primary transition region   83
    6.1   Mechanical relaxation   83
    6.2   Dielectric relaxation   90
    6.3   Mechanical vs. dielectric relaxation   96
    6.4   NMR relaxation   104
    6.5   Neutron scattering   110
          References   118
7   Secondary (subglass) relaxations   120
    7.1   Occurrence of mechanical and dielectric secondary processes   120
    7.2   Complexity and multiplicity of secondary processes   121
    7.3   Flexible side group motion as a source of secondary relaxation   129
    7.4   NMR spectroscopy studies of flexible side group motion   138
          References   140
8   The transition from melt to glass and its molecular basis   142
    8.1   Experimental description   142
    8.2   Molecular basis   157
          References   194
Part III   Complex systems   197
9   Semi-crystalline polymers   199
    9.1   Phase assignment   200

9.2   Effect of crystal phase presence on amorphous
        fraction relaxation   209
9.3   Relaxations in semi-crystalline polymers with a crystal
        phase relaxation   214
9.4   NMR insights   223
        References   226
10   Miscible polymer blends   227
10.1   Poly(isoprene)/poly(vinyl ethylene) (PI/PVE) blends   228
10.2   Models for miscible blend dynamics   229
10.3   MD simulations of model miscible blends   233
10.4   PI/PVE blends revisited   239
        References   243
Appendix AI   The Rouse model   244
AI.1   Formulation and normal modes   244
AI.2   Establishment of Rouse parameters for a real polymer   245
AI.3   The viscoelastic response of a Rouse chain   245
AI.4   Bead displacements and the coherent single-chain
        structure factor   246
        References   247
Appendix AII   Site models for localized relaxation   248
AII.1   Dipolar relaxation in terms of site models   248
AII.2   Mechanical relaxation in terms of site models   251
        References   252
   Index   253

Preface

Polymers have become widely used materials because they exhibit an enormous range of behaviors and properties. They are most often processed or shaped as viscous melts. They can be used as stiff solid materials in the glassy or semicrystalline state. The rubbery or elastomeric state, obtained by cross-linking melts, is characterized by very high reversible extensibility and is unique to polymeric molecular organization. In addition, many applications are dependent upon the exhibition of behavior intermediate between that of the viscous melt and that of the relatively rigid glassy state. That is, the degree of rigidity is time dependent. In the solid state, major changes in physical properties can occur with changing temperature. Thus the same polymer can be a melt or, if cross-linked, an elastomer, or a somewhat rigid glass, or a quite rigid glass depending on the time and temperature of use. Further, these changes of properties occur in regions of time and temperature that are well defined. That is, the regions can be characterized by a variety of experimental techniques that probe the relaxation of the response following an applied perturbation such as mechanical stress or an electric field. Most polymers exhibit several such relaxation regions.

All of this rich manifold of behavior has its foundation in the ability of polymer molecules to locally change the details of the shape or conformation of the molecular chain and to accumulate these changes so that global changes in molecular shape can result. These local changes usually involve rotations about the constituent bonds and attendant responses of nearby bonds and surrounding chains. Energy barriers are involved and thus thermally activated responses result. This then leads to the time and temperature dependent character of a relaxation process. If relaxation processes are to be fully understood, then these molecular scale events have to be understood as well.

Four experimental methods are considered. Historically, perhaps the greatest early interest in relaxation phenomena centered on mechanical response as exemplified by creep and stress relaxation experiments. Concurrently, however, and probably because mechanical experiments have often been considered difficult to carry out over broad ranges of time or frequency, dielectric response measurements where the short time, high frequency region is more conveniently accessed became popular. A very large literature has developed around both the mechanical and dielectric response methods. Somewhat later it was found that the decay of polarization of nuclear spins associated with the nuclear magnetic resonance (NMR) method was sensitive to motional processes and could be invoked as a tool for relaxation studies. The specificity to certain atoms in particular bonding environments is an advantage. That and the development of pulse techniques that allow wide ranges of time to be explored have led to increasingly important applications to polymer relaxations. Several scattering processes, including Rayleigh and Brillouin scattering of light and neutron scattering, are also sensitive to motional processes. But, of these, only neutron diffraction is considered here.

Experiments rarely give direct insight into the details of the molecular motions underlying relaxation processes. However, by rationalizing the results from several experimental techniques applied to groups of structurally similar but distinct polymers a reasonable mechanism can often be formulated. The process of molecular mechanism elucidation has been significantly aided by the advent of computer assisted detailed atomistic molecular modeling. Particularly valuable is the molecular dynamics method which gives the positions of every atom as a function of time. From this information time autocorrelation functions (ACF) can be constructed utilizing linear response theory that can be compared with experimental data for various techniques. The use of simulation as a tool in molecular interpretation is heavily stressed.

Part I

Methodology

The first five chapters deal with the methodology used to study relaxation processes. The first four deal with experimental methods, namely mechanical relaxation, dielectric relaxation, nuclear magnetic resonance and neutron diffraction. The first two are very familiar long-used methods. This is due both to their relevance to practical material properties and to the insights they have led to in understanding the time dependence of material behavior. The NMR method is very selective in probing certain elements based, of course, on the nuclear spins involved. The development of complex pulse techniques has allowed the investigation of relaxational processes over very broad time scales. The availability of sophisticated neutron sources has led to a wealth of data involving dynamic scattering factors for coherent and incoherent scattering. Molecular dynamics (MD) simulations are becoming more and more useful in the interpretation of the various relaxation experiments and Chapter 5 presents the basics of that methodology.

1

Mechanical relaxation

1.1 Regimes of behavior

Amorphous polymers tend to behave in an elastic manner at low temperature in the glassy state. The strain at break is usually small (a few percent), they can deform quickly, hold their length at constant load, and recover completely when unloaded (Figure 1.1). The material is *elastic*. In the vicinity of the glass transition temperature when the length of the specimen is held constant the stress decays with time (stress relaxation). Under constant load (creep), in addition to the instantaneous deformation characteristic of the glass, the sample deforms in a time dependent fashion and when released from constant load recovers nearly completely and in a time dependent manner (retarded elasticity) (Figure 1.2). Such a material is called *anelastic*. At higher temperature, in addition to the instantaneous and retarded elasticity a non-recoverable strain appears in the creep experiment due to viscous flow (Figure 1.3). The material is *viscoelastic*.

There is a considerable advantage in being able to describe and summarize the above behavior in terms of a simple model. For example, it would be impractical to perform every type of mechanical test on a sample. Rather it would be much better to perform enough measurements to characterize the material and then predict the results of other tests from a model. Continuum linear elasticity theory is soundly based on the application of classical mechanics to the deformation of solids. However, in order to include time dependent material behavior an independent additional conjecture about how such materials behave must be introduced. Because the conjecture has no obvious molecular or other fundamental derivation the model is said to be a *phenomenological* one. Based on the above comments it is apparent that a first step is to introduce the conjecture upon which linear viscoelastic behavior is based. This is the *superposition* assumption.

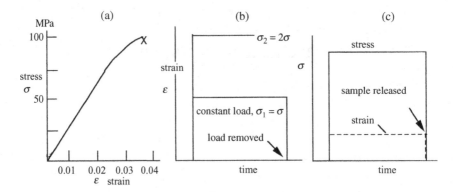

Figure 1.1 Glassy state: (a) stress–strain, (b) creep and (c) stress relaxation curves. For small strains, strain is proportional to stress (see (b)). Recovery from load or strain is complete and rapid. The material is *elastic*.

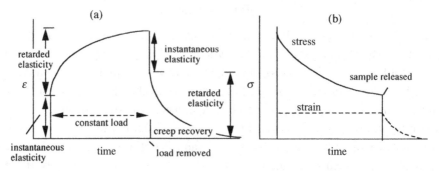

Figure 1.2 Near the glass temperature: (a) in the creep curve there is, in addition to the instantaneous strain of the glassy state, a time dependent strain. It is largely recoverable on removal of the stress. (b) In stress relaxation, the stress decays with time. The material is said to be *anelastic*.

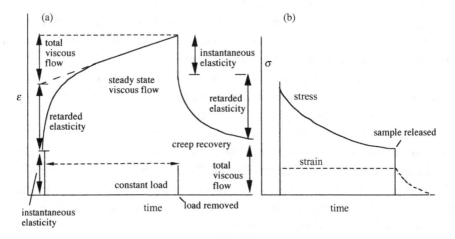

Figure 1.3 Well above glass temperature: (a) creep and (b) stress relaxation curves. In addition to instantaneous and retarded elasticity, a non-recoverable component is important, i.e. viscous flow. The material is *viscoelastic*.

1.2 Superposition principle

In linear continuum elasticity theory, *linear* means that stress and strain are proportional *and* that strains occasioned by multiaxial stresses and vice versa are additive or superposable (see Appendix A1). In linear viscoelasticity the same linearity is invoked. But in addition and of more significance in the present context is the (Boltzmann [1]) *superposition principle* that describes the time dependence. The central consequence of time dependent behavior is the necessity for the concept of past time *and* current time. In creep, for example (Figure 1.1), when a perfectly elastic material is unloaded, the strain immediately responds. Thus there is only one concept of time necessary, the record of the loading schedule. In Figure 1.2 where time dependence does appear, a plot with only one time axis can still be made. This is because a simple loading history, the sudden application or removal of the load, was invoked. Suppose, however, the load in creep varies slowly over time. For an elastic material this is of no consequence since the strain responds quickly (subject only to inertia). For an anelastic or viscoelastic material this is no longer the case. The strain at the *current* time is still changing from *past* applications of the load. Thus an additional assumption about material behavior is required.

The superposition principle states that the time evolution of a strain response to a past stress is independent of any stresses applied in the intervening period up to the time at which the strain is currently being measured. Thus the present time strain is the *sum* of all strains arising from past applications of stress. The converse applies when the strain history is imposed and stress occurs in response (stress relaxation). The present time stress is the sum of all stresses arising from past applications of strain.

In what follows all of the examples are given in terms of uniaxial tensile stress–strain measurements. Similar formulations and equations follow for shear measurements.

1.3 Relaxation modulus

The superposition principle is easily mathematically quantified. The case of an imposed strain history is used in illustration. The time dependence is embedded in a time dependent modulus that is characteristic of the material. This modulus is a function of the elapsed time between the application of an imposed strain at past time, u, and the current time, t, at which the stress is being measured. The modulus, now called a *relaxation modulus*, is denoted as $E_R(t - u)$. Suppose a strain increment, $\Delta\varepsilon(u)$ is applied at time u. Then the stress increment at later time, t, arising from this will be

$$\Delta\sigma(t) = E_R(t - u)\Delta\varepsilon(u). \tag{1.1}$$

The total stress at the current time, t, from a series of past strain applications at times $u_1, u_2, \ldots, u_i, \ldots$ is, under the superposition principle,

$$\sigma(t) = E_R(t - u_1)\Delta\varepsilon(u_1) + E_R(t - u_2)\Delta\varepsilon(u_2)$$
$$+ \cdots + E_R(t - u_i)\Delta\varepsilon(u_i) + \cdots. \tag{1.2}$$

An arbitrary strain history can be built up by passing to the integral,

$$\sigma(t) = \int_{-\infty}^{t} E_R(t - u)(d\varepsilon/du)\,du, \tag{1.3}$$

where the device of introducing $d\varepsilon/du$ enables the integration to be carried out over past time, u. This equation forms the basis for the description of the stress response under all strain histories. In other words, everything about the material is contained in $E_R(t - u)$, everything about various possible experiments, tests, etc. is contained in $\varepsilon(u)$. It is obviously presumed that the relaxation modulus $E_R(t - u)$ is known. The strategy is to determine it experimentally under some standard imposed strain history. Then the results of other strain histories may be predicted from E_R using eq. (1.3).

As a mathematical convenience later, it will be useful to explicitly recognize that at long times the stress generally decays not to zero but to a constant value expressed through the equilibrium fully *relaxed modulus*, E_r. Thus eq. (1.3) becomes

$$\sigma(t) = E_r\varepsilon(t) + \int_{-\infty}^{t} (E_R(t - u) - E_r)(d\varepsilon/du)\,du. \tag{1.4}$$

The term $(E_R(t - u) - E_r)$ does approach zero at large values of $t - u$.

1.4 Simple stress relaxation

A common experiment is to impose a sudden finite strain step at $u = 0$ that remains constant at ε^0 and to observe the ensuing stress as a function of time. Under this circumstance eq. (1.3) or eq. (1.4) integrates to

$$\sigma(t) = E_R(t)\varepsilon^0. \tag{1.5}$$

Thus the relaxation modulus is directly determined by the experiment (for times greater than several times the rise time of the step strain) as $E_R(t) = \sigma(t)/\varepsilon^0$.

1.5 Dynamic modulus

A common experiment is the imposition of a cyclic stress or strain. It is a mathematical convenience to express such periodic functions in complex notation. Thus an imposed periodic strain of angular frequency ω can be written as

$$\varepsilon^*(i\omega t) = \varepsilon^0 e^{i\omega t} = \varepsilon^0[\cos(\omega t) + i\sin(\omega t)]. \tag{1.6}$$

It is implied that the actual strain is the real part of $\varepsilon^*(i\omega\tau)$, or, therefore $\varepsilon^0\cos(\omega t)$. Since eq. (1.4) is linear in ε, σ may be found from the real part of σ^* in

$$\sigma^*(i\omega t) = E_r\varepsilon^0 e^{i\omega t} + \int_{-\infty}^{t} (E_R(t-u') - E_r)(i\omega\varepsilon^0 e^{i\omega u'})\,du', \tag{1.7}$$

where past time is labeled u' and where eq. (1.6) has been used to find $d\varepsilon/du'$. Making the substitution, $u = t - u'$ gives

$$\sigma^*(i\omega t) = \varepsilon^0 e^{i\omega t}\left[E_r + \int_0^{\infty}(E_R(u) - E_r)(i\omega e^{-i\omega u})\,du\right]. \tag{1.8}$$

It is convenient to define a *complex dynamic modulus* $E^*(i\omega)$ such that

$$\sigma^*(i\omega t) = E^*(i\omega)\varepsilon^*(i\omega t), \tag{1.9}$$

where

$$E^*(i\omega) = E_r + i\omega \int_0^{\infty}(E_R(u) - E_r)e^{-i\omega u}\,du. \tag{1.10}$$

Although only the real part of $\sigma^*(i\omega t)$ is of physical significance, since the right hand side of eq. (1.9) involves the product of two complex numbers, the imaginary part of $E^*(i\omega)$ does have physical meaning as it gives rise to a real term in $\sigma^*(i\omega t)$. The real and imaginary parts of $E^*(i\omega)$ in

$$E^*(i\omega) = E'(\omega) + iE''(\omega) \tag{1.11}$$

are given by

$$E'(\omega) = E_r + \omega \int_0^{\infty}(E_R(u) - E_r)\sin(\omega u)\,du, \tag{1.12}$$

$$E''(\omega) = \omega \int_0^{\infty}(E_R(u) - E_r)\cos(\omega u)\,du. \tag{1.13}$$

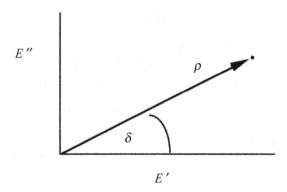

Figure 1.4 Complex plane representation of E^*.

The physical significances of E' and E'' are best seen by writing E^* in polar form (see Figure 1.4) as

$$E^* = \rho e^{i\delta},\tag{1.14}$$

where

$$\rho = (E'^2 + E''^2)^{1/2}\tag{1.15}$$

and

$$\tan \delta = E''/E'.\tag{1.16}$$

Thus,

$$\sigma^*(i\omega t) = \rho\varepsilon^0 e^{i(\omega t + \delta)}\tag{1.17}$$

and

$$\sigma(t) = \rho\varepsilon^0 \cos(\omega t + \delta) = \sigma^0 \cos(\omega t + \delta).\tag{1.18}$$

Therefore it may be seen that the stress in response to an imposed periodic strain of angular frequency ρ is also periodic with the same frequency but leads the strain by an angle δ. See Figure 1.5.

The maximum stress, i.e., its amplitude, is given by $\sigma^0 = \rho\varepsilon^0$. The real component of E^*, E', is called the *dynamic storage modulus* and the imaginary part, E'', is called the *dynamic loss modulus*. The reasons for these designations are the following.

The work done per unit volume, W, in deforming the material by a small strain is given by

$$dW = \sigma d\varepsilon = \text{Re}(E^*\varepsilon^*)\,d\varepsilon.\tag{1.19}$$

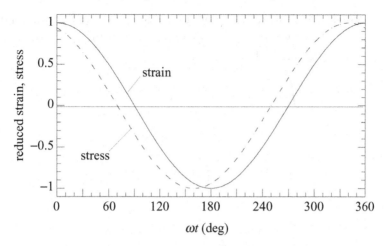

Figure 1.5 The stress and strain over one cycle in a dynamic experiment. Both are normalized to unity as, $\varepsilon/\varepsilon^0$ and σ/σ^0. Phase angle $\delta = 20°$.

The work associated with E' is reversible and is zero over a complete cycle. However, as seen below E'' leads to energy absorption. The work per unit volume per second at frequency, v, is

$$\dot{W} = \mathrm{d}W/\mathrm{d}t = v\mathrm{Re}\left[\int_0^{t=1/v} E^*\varepsilon^*(\mathrm{d}\varepsilon/\mathrm{d}u)\,\mathrm{d}u\right] \tag{1.20}$$

$$= \frac{\omega}{2\pi}\varepsilon^{0\,2}\mathrm{Re}\left[\int_0^{2\pi/\omega}(E' + iE'')e^{i\omega u}(e^{i\omega u} - e^{-i\omega u})\frac{i\omega}{2}\,\mathrm{d}u\right] \tag{1.21}$$

$$= \omega E''\varepsilon^{0\,2}/2. \tag{1.22}$$

1.6 Interconversion of stress relaxation and dynamic modulus

The complex dynamic modulus may be measured directly via the stress magnitude and the phase angle in response to an applied periodic strain of magnitude ε^0. However, it may also be derived indirectly from the results of a stress relaxation experiment using eq. (1.10) or equivalently eq. (1.12) and eq. (1.13). Notice, however, that the limits in these integrations are 0, ∞. The *entire* stress relaxation function over time must in principle be known. Because of the rather slowly varying with time or frequency nature of most polymeric relaxations this latter condition can be relaxed considerably. More on this is introduced later.

It is also possible to convert the dynamic modulus into the relaxation modulus. This is done by invoking the Fourier transformation properties associated with eq. (1.10) or eq. (1.12), eq. (1.13). That is, if

$$f(x) = \sqrt{\frac{2}{\pi}} \int_0^\infty F_c(y) \cos(xy) dy = \sqrt{\frac{2}{\pi}} \int_0^\infty F_s(y) \sin(xy) dy, \qquad (1.23)$$

then

$$F_c(y) = \sqrt{\frac{2}{\pi}} \int_0^\infty f(x) \cos(xy) dx \qquad (1.24)$$

and

$$F_s(y) = \sqrt{\frac{2}{\pi}} \int_0^\infty f(x) \sin(xy) dx, \qquad (1.25)$$

from which it follows that

$$E_R(t) = E_r + \frac{2}{\pi} \int_0^\infty [E'(\omega) - E_r] \sin(\omega t) d\omega/\omega \qquad (1.26)$$

or

$$E_R(t) = E_r + \frac{2}{\pi} \int_0^\infty E''(\omega) \cos(\omega t) d\omega/\omega. \qquad (1.27)$$

Thus interconversion of stress relaxation and dynamic data is possible via numerical Fourier transform methods provided the data cover or can be accurately extrapolated over the entire time or frequency range.

Since both $E'(\omega)$ and $E''(\omega)$ may be expressed in terms of a single function, $E_R(t)$, it is apparent that $E'(\omega)$ and $E''(\omega)$ must be themselves related. A relationship may be found by substituting eq. (1.27) into eq. (1.12),

$$E'(\omega) = E_e + \omega \int_0^\infty \frac{2}{\pi} \left[\int_0^\infty E''(x) \cos(xu)(dx/x) \right] \sin(\omega u) du \qquad (1.28)$$

$$= E_e + \frac{2}{\pi}\omega \lim_{R \to \infty} \int_0^\infty \left[\int_0^R E''(x) \cos(xu) \sin(\omega u) du(dx/x) \right] \qquad (1.29)$$

$$= E_e + \frac{2\omega}{\pi} \lim_{R \to \infty} \int_0^\infty E''(x) \left[\frac{1-\cos(\omega-x)R}{2(\omega-x)} + \frac{1-\cos(\omega+x)R}{2(\omega+x)} \right] (dx/x)$$

$$(1.30)$$

and since the integrals involving the cosine terms vanish in the limit of large R it follows that

$$E'(\omega) = E_e + \frac{2\omega^2}{\pi} \int_0^\infty \frac{E''(x)}{(\omega^2 - x^2)} d\ln x. \tag{1.31}$$

This is known as a *Kramers–Kroenig relation* [2,3]. In addition to expressing the relationship between $E''(\omega)$ and $E'(\omega)$ at all frequencies, eq. (1.31) leads to a useful specific relation concerning the entire relaxation process. As ω becomes very high, the modulus approaches a high frequency *unrelaxed* value E_u and therefore,

$$E_u = E'(\infty) = E_r + \frac{2}{\pi} \int_0^\infty E''(x) d\ln x. \tag{1.32}$$

In other words, *the increment in the modulus, $E_u - E_r$ over the relaxation is equal to $2/\pi$ times the area under the loss modulus curve*, when the latter is plotted against $\ln \omega$.

1.7 Representation of the relaxation function: single relaxation time (SRT)

Up to this point the nature of the relaxation function, i.e., the relaxation modulus, in eq. (1.3) or eq. (1.4) has been left arbitrary. The simplest mathematical description of a function that relaxes monotonically from an initial value, E_u, to a long-time equilibrium value, E_r, would use an exponential dependence on time or

$$E_R(t) = (E_u - E_r)e^{-t/\tau} + E_r, \tag{1.33}$$

where τ is the *relaxation time*. This is the single relaxation time (SRT) representation. Use of eq. (1.33) in eq. (1.10) gives the following relation:

$$E^*(i\omega) = E_r + (E_u - E_r)i\omega\tau/(1 + i\omega\tau). \tag{1.34}$$

Upon rationalization into real and imaginary parts, the storage and loss components of the dynamic modulus are found as

$$E'(\omega) = E_r + (E_u - E_r)\omega^2\tau^2/(1 + \omega^2\tau^2), \tag{1.35}$$
$$E''(\omega) = (E_u - E_r)\omega\tau/(1 + \omega^2\tau^2). \tag{1.36}$$

The behavior of $E_R(t)$ is shown in Figure 1.6 and can be compared with that of $E'(\omega)$, $E''(\omega)$ in Figure 1.7. Notice in Figure 1.7 that the storage modulus rises monotonically from the equilibrium E_r value to the unrelaxed, E_u value as frequency increases. In contrast the loss modulus has a maximum value at $\omega\tau = 1$. The appearance of the maximum in the latter suggests that dynamic relaxation tests

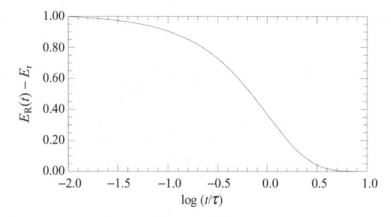

Figure 1.6 Relaxation modulus (unit strength, $E_u - E_r$) vs. log time for the SRT representation.

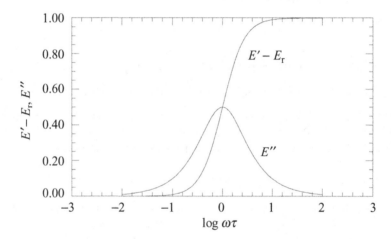

Figure 1.7 Dynamic storage modulus and loss vs. log (circular) frequency for the (unit strength) SRT representation.

have the *attribute of being a form of spectroscopy*. That is, peaks in the loss modulus appear as frequency is scanned.

The increment in the modulus may be regarded as the *strength* of the relaxation as it expresses the change in the property as a result of the entire relaxation process. The concept of relaxation *peak height* is manifested in $E''(\omega)$ vs. ln ω. Notice from eq. (1.32), however, it is the *area* under the loss curve that determines relaxation strength and not peak height per se.

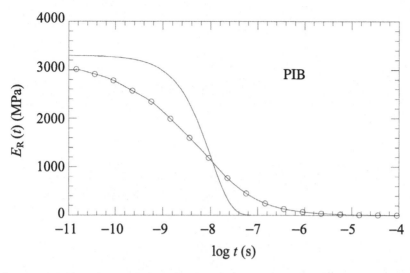

Figure 1.8 Relaxation modulus of PIB plotted against log t (open circles). Also shown for comparison is SRT behavior (solid curve no points).

1.8 Relaxations in polymeric materials tend to be "broad"

It is a matter of experimental observation that the modulus relaxation behavior of an actual polymer deviates, and usually quite seriously, from SRT behavior. This is illustrated in Figure 1.8, where the relaxation modulus of poly(isobutylene) (PIB) [4] at 25 °C is plotted on a logarithmic time scale. Also shown is the SRT response, eq. (1.33), on a logarithmic time basis and shifted on the time axis by the choice of τ to correspond to the relaxation region for PIB. It is apparent that the observed behavior covers many orders of magnitude in time in contrast to approximately 2 for the SRT representation.

A feature of the stress relaxation experiment above is that the modulus can be measured to extremely low values in comparison with the starting modulus. In fact in the glass transition region in amorphous polymers the modulus decays to values orders of magnitude below that of the unrelaxed initial modulus. Thus a logarithmic scale is more appropriate for the presentation of the modulus. Figure 1.9 shows the PIB relaxation data of Figure 1.8 plotted with both of the axes as logarithmic scales. It may now be seen that there are additional features in the relaxation behavior. One is that the glass transition region relaxes to a highly entangled elastic melt and the other is the onset at long time of viscous flow in the entangled melt. In the viscous flow region, stress is proportional to shear rate. Thus for a constant rate the stress remains constant but strain continues to grow indefinitely and the apparent modulus decreases indefinitely.

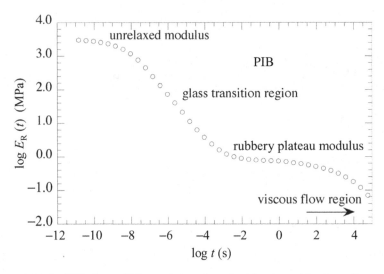

Figure 1.9 The PIB data of Figure 1.8 replotted with a logarithmic ordinate and extended to longer time.

1.9 Distribution of relaxation times

In an empirical sense the SRT behavior can be generalized to represent arbitrary behavior by supposing that there are a number of pathways by which partial relaxation can take place, each exhibiting simple exponential behavior. Alternatively, simply as a mathematical expansion, a number of exponential terms can be invoked. In any case the relaxation modulus can be written as

$$E_R(t) = E_r + \Delta E_1 \exp(-t/\tau_1) + \Delta E_2 \exp(-t/\tau_2) + \Delta E_3 \exp(-t/\tau_3) + \cdots \tag{1.37}$$

and therefore

$$E_R(i\omega) = E_r + \Delta E_1 i\omega\tau_1/(1 + i\omega\tau_1) + \Delta E_2 i\omega\tau_2/(1 + i\omega\tau_2)$$
$$+ \Delta E_3 i\omega\tau_3/(1 + i\omega\tau_3) + \cdots. \tag{1.38}$$

In the limit of many closely spaced relaxation times, the finite spectrum approaches a continuous one and

$$E_R(t) = E_r + \int_{-\infty}^{+\infty} \overline{H}(\tau) e^{-t/\tau} \, d\ln\tau, \tag{1.39}$$

$$E^*(i\omega) = E_r + \int_{-\infty}^{+\infty} \overline{H}(\tau) \frac{i\omega\tau}{1 + i\omega\tau} \, d\ln\tau, \tag{1.40}$$

where $\overline{H}(\tau)$ is the *spectrum of* (logarithmic) *relaxation times*. Because $\overline{H}(\tau)$ is an arbitrary function it is presumed that any behavior of the relaxation and dynamic moduli could be accommodated by a proper choice of the function.

1.10 Relaxation spectrum from $E_R(t)$

Unlike the relations between $E_R(t)$ and $E^*(i\omega)$, where either can be expressed in terms of the other by Fourier transformation, the spectrum in eq. (1.39) and eq. (1.40) is not readily obtained by inversion. Numerical methods are resorted to and these have been much researched (see, for example, Ferry [4] and also Alvarez *et al.* [5] and MacDonald [6]). Just one example is given here [7]. It is convenient, robust, and also illustrates the ameliorating conditions found in polymer relaxation data that permit transformations of data without fulfilling the rigorous condition of requiring the entire time or frequency spectrum to do so. The example is based on the fact that as a function of $\log t$ the relaxation modulus, $E_R(t)$ is often slowly varying.

To start, it is assumed that the experimental data have been fit with a cubic spline. This can be an unadorned interpolating spline that passes through each data point and has continuous first and second derivatives or a "smoothing" spline that minimizes the residual for a group of points. In any event the interpolated data and the first and second derivatives are available at any point. The method is based on the premise that the distribution function is such a slowly varying function of τ that $e^{-t/\tau}$, regarded as a function of τ, changes rapidly in comparison with the distribution function.

For temporary convenience a distribution in linear time, $\overline{K}(\tau)$, such that $\overline{K}(\tau)\,d\tau = \overline{H}(\tau)\,d\ln\tau$ is invoked. In the ensuing equation for the relaxation modulus,

$$E_R(t) = \int_0^\infty \overline{K}(\tau)e^{-t/\tau}\,d\tau, \tag{1.41}$$

the exponential, $e^{-t/\tau}$ is replaced by the approximation (see Figure 1.10),

$$e^{-t/\tau} = \begin{cases} 0, & \tau < t \\ 1, & \tau \geq t \end{cases} \tag{1.42}$$

so that,

$$E_R(t) = \int_{\tau=t}^\infty \overline{K}(\tau)\,d\tau. \tag{1.43}$$

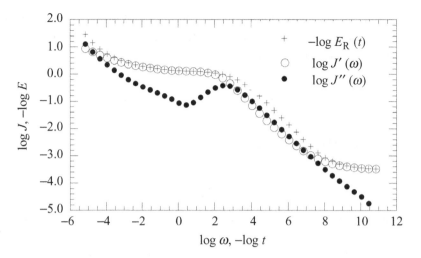

Figure 1.13 Dynamic compliance of PIB calculated from the dynamic modulus results in Figure 1.12. The relaxation modulus $E_R(t)$ is also shown for comparison.

and

$$J'' = E''/(E'^2 + E''^2).\qquad(1.60)$$

Notice that

$$\tan\delta = J''/J' = E''/E'.\qquad(1.61)$$

From eq. (1.59) and eq. (1.60) it is apparent that conversion of dynamic modulus data to dynamic compliance data and vice versa is trivial and requires only a minor arithmetic calculation at each datum point frequency. The conversion of the dynamic modulus data for PIB in Figure 1.12 to dynamic compliance is illustrated in Figure 1.13.

Finally, the creep curve, $J(t)$, can be constructed from the dynamic creep response, $J^*(\omega)$ via Fourier transformation. The analogs of eq. (1.26) and eq. (1.27) applied to creep are

$$J(t) = J_u + \frac{2}{\pi}\int_0^\infty [J'(\omega) - J_u]\sin(\omega t)d\omega/\omega,\qquad(1.62)$$

$$J(t) = J_u + \frac{2}{\pi}\int_0^\infty J''(\omega)\cos(\omega t)d\omega/\omega.\qquad(1.63)$$

The result of the use of eq. (1.62) to compute $J(t)$ is displayed in Figure 1.14.

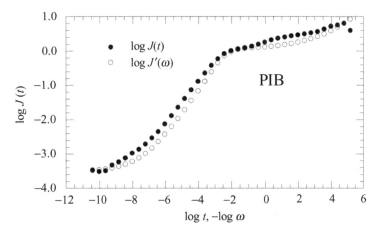

Figure 1.14 Creep curve $J(t)$ for PIB computed from the dynamic creep $J^*(\omega)$ real component $J'(\omega)$.

1.13 Representation of the retardation function

The counterpart of the SRT function in eq. (1.33) is the *single retardation time function*,

$$J(t) = (J_r - J_u)(1 - e^{-t/\tau}) + J_u. \tag{1.64}$$

Notice that this function has an initial unretarded value of J_u and increases to an equilibrium unretarded relaxed value of J_r. When eq. (1.64) is employed in eq. (1.56) the complex dynamic compliance is found as,

$$J^*(i\omega) = J_u + (J_r - J_u)/(1 + i\omega\tau). \tag{1.65}$$

When rationalized the real and imaginary components are found as

$$J' = J_u + (J_r - J_u)/(1 + \omega^2\tau^2), \tag{1.66}$$

$$J'' = (J_r - J_u)\omega\tau/(1 + \omega^2\tau^2). \tag{1.67}$$

As noted above, the convention $J^* = J' - iJ''$ is used in order that J'' be positive. The dynamic compliance components, J', J'', for the SRT representation are shown in Figure 1.15.

As in eq. (1.39) and eq. (1.40) for the modulus, the single retardation time representation can be generalized to include a *spectrum of retardation times*, $\overline{L}(\tau)$:

$$J(t) = J_u + \int_{\ln\tau=-\infty}^{+\infty} \overline{L}(\tau)(1 - e^{-t/\tau})\,d\ln\tau, \tag{1.68}$$

$$J^*(i\omega) = J_u + \int_{\ln\tau=-\infty}^{+\infty} \overline{L}(\tau)\left(\frac{1}{1 + i\omega\tau}\right)d\ln\tau. \tag{1.69}$$

where a high frequency "unrelaxed" limiting relative permittivity, ε_u, has also been inserted.

A complex dynamic relative permittivity, $\varepsilon^*(i\omega)$, is defined as

$$D(i\omega t) = \varepsilon^*(i\omega)E(i\omega t) \tag{2.24}$$

and therefore (cf. eq. (1.56)),

$$\varepsilon^*(i\omega) = \varepsilon_u + i\omega \int_0^\infty (\varepsilon(u) - \varepsilon_u)e^{-i\omega u}\,du. \tag{2.25}$$

2.4.2 Relationship to measurements

From eq. (2.22) it is apparent that time domain measurement of $\varepsilon(t)$ can be effected in a straightforward manner from the time dependence of the charging of a capacitor in response to a step voltage. In the dynamic case some elaboration is appropriate. It is convenient to suppose that the complex impedance, $Z(i\omega)$, of a capacitor containing the dielectric has been measured at the frequency ω. The expressions for the permittivity in eq. (2.16) and eq. (2.17) can be generalized by replacing the real capacitance C by its complex impedance equivalent $1/i\omega Z(i\omega)$, or

$$\varepsilon^*(i\omega) = \frac{1}{i\omega Z(i\omega)}\left(\frac{d}{\kappa_0 A}\right) \tag{2.26}$$

$$= \frac{1}{i\omega Z(i\omega)C_0}. \tag{2.27}$$

The measuring instrument often expresses the measured impedance in terms of an equivalent circuit for the sample filled capacitor. The most commonly used equivalent circuit considers the sample to be a capacitor and resistor connected in parallel (Figure 2.7). In terms of this circuit the impedance is given by

$$\frac{1}{Z(i\omega)} = i\omega C + \frac{1}{R} \tag{2.28}$$

from which it follows by reference to eq. (2.26) and eq. (2.27) that

$$\varepsilon' = \frac{Cd}{\kappa_0 A} = C/C_0, \tag{2.29}$$

$$\varepsilon'' = \frac{1}{\omega R}\left(\frac{d}{\kappa_0 A}\right) = \frac{1}{\omega RC_0}, \tag{2.30}$$

where R and C are the values furnished by the measuring instrument. The latter are understood to be frequency dependent in the sense that values from the instrument respond to the frequency dependence of the complex permittivity, $\varepsilon^*(i\omega)$.

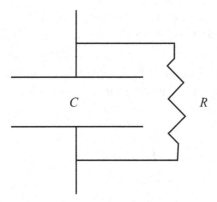

Figure 2.7 Parallel capacitor–resistor equivalent circuit of a sample.

2.4.3 Direct current conduction

The equivalent circuit in Figure 2.7 can also have direct physical significance. If the specimen exhibits steady state charge migration (DC conductance) the resistance R in Figure 2.7 can be considered to be independent of frequency. In that case the conductance will give rise to an apparent loss, ε'', represented by eq. (2.30). It is to be noted that loss from this origin increases without limit as frequency decreases. This can have a deleterious effect in studies of low frequency, dipolar reorientation processes in that such processes can be masked by loss from low frequency conductance loss.

2.5 Empirical representations of the dielectric permittivity

2.5.1 SRT

Representation of single retardation time (τ) behavior for the dielectric case involves only transcription of the elastic compliance results of Sec. 1.12:

$$\varepsilon(t) = (\varepsilon_r - \varepsilon_u)(1 - e^{-t/\tau}) + \varepsilon_u. \tag{2.31}$$

The function has an initial unrelaxed value of ε_u and increases to a relaxed long time value of ε_r. The complex dynamic relative permittivity is found as

$$\varepsilon^*(i\omega) = \varepsilon_u + (\varepsilon_r - \varepsilon_u)/(1 + i\omega\tau). \tag{2.32}$$

The real (storage) and imaginary (loss) components are:

$$\varepsilon' = \varepsilon_u + (\varepsilon_r - \varepsilon_u)/(1 + \omega^2\tau^2), \tag{2.33}$$
$$\varepsilon'' = (\varepsilon_r - \varepsilon_u)\omega\tau/(1 + \omega^2\tau^2). \tag{2.34}$$

The convention $\varepsilon^* = \varepsilon' - i\varepsilon''$ is again used in order that ε'' be positive.

In the dielectric relaxation context, eqs. (2.32)–(2.34) are known as the *Debye* equations as they were first derived by him from a model formulated to represent dipolar relaxation in simple liquids [1, 2]. The dielectric permittivity components, ε', ε'', for the SRT representation are the same as shown in Figure 1.14.

As in Sec.1.12, the single retardation time representation can be generalized to include a *spectrum of retardation times, $\overline{F}(\tau)$*:

$$\varepsilon(t) = \varepsilon_u + \int_{\ln \tau = -\alpha}^{+\infty} \overline{F}(\tau)(1 - e^{-t/\tau}) \, d\ln \tau, \tag{2.35}$$

$$\varepsilon^*(i\omega) = \varepsilon_u + \int_{\ln \tau = -\alpha}^{+\infty} \overline{F}(\tau) \left(\frac{1}{1 + i\omega\tau} \right) d\ln \tau. \tag{2.36}$$

2.5.2 The stretched exponential, "KWW", time domain function

As noted in Chapter 1 the SRT representation is inadequate in that observed relaxation processes in polymers are usually much broader. In dielectric relaxation work it has been traditional to represent the experimental behavior by means of empirical equations that fit the data well. This procedure is permitted by the relatively broad time or frequency span accessible in the dielectric method. The experimental data can then be summarized by a few parameters that appear in the equations. For data expressed in the time domain a simple empirical modification of eq. (2.31) has proven effective and popular in application to the glass transition region. The t/τ term in the exponential is replaced by $(t/\tau)^{\beta}$ where β is an empirical parameter. Thus,

$$\varepsilon(t) = \varepsilon_u + (\varepsilon_r - \varepsilon_u)\left(1 - e^{-(t/\tau)^{\beta}}\right). \tag{2.37}$$

Values of $\beta < 1$ introduce broadening in the time response. The equation was apparently first used by Kohlrausch in an unrelated context [3]. In statistics it is known as the Weibull distribution. In polymer behavior it has been used to represent mechanical compliance [4, 5, 6]. In more recent times it has been used by Williams and Watts [7] for the representation of time domain dielectric data, hence the appellation Kohlrausch–Williams–Watts or more briefly as "KWW." Figure 2.8 shows time domain data for poly(vinyl acetate) (PVAc) in the glass transition region fitted with the KWW function. Although systematic discrepancies in the fit can be discerned the function does a very creditable job of representing the data.

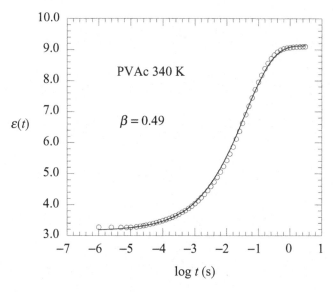

Figure 2.8   Time domain dielectric permittivity for PVAc in the glass transition region along with a KWW stretched exponential fit to the data. The data are from measurements reported by Boyd and Liu [8].

2.5.3 Frequency domain representations

The KWW time domain function is an empirical extension of the SRT time function, eq. (2.31). Similarly, the frequency domain representation of the SRT, eq. (2.32), can be empirically modified. Historically, the first modification was introduced by Cole and Cole [9] in the following manner. The $i\omega\tau$ term is replaced by $(i\omega\tau_0)^\alpha$, where α is a broadening parameter (for $\alpha < 1$) and the subscript on τ is introduced to indicate a *central relaxation time*. Thus,

$$\text{Cole–Cole}\quad \varepsilon^*(i\omega) = \varepsilon_u + (\varepsilon_r - \varepsilon_u)/[1 + (i\omega\tau_0)^\alpha]. \tag{2.38}$$

This is known as the *Cole–Cole function* and it is much used in dielectric data fitting. Rationalization of eq. (2.38) into the real and imaginary components by making the substitution, $i = e^{i\pi/2}$ in the denominator and multiplying numerator and denominator by the complex conjugate of the denominator which has $e^{-i\pi/2}$ for i, leads to

$$\varepsilon' = \varepsilon_u + (\varepsilon_r - \varepsilon_u) \left(\frac{1 + (\omega\tau_0)^\alpha \cos\left(\frac{\pi}{2}\alpha\right)}{1 + 2(\omega\tau_0)^\alpha \cos\left(\frac{\pi}{2}\alpha\right) + (\omega\tau_0)^{2\alpha}} \right) \tag{2.39}$$

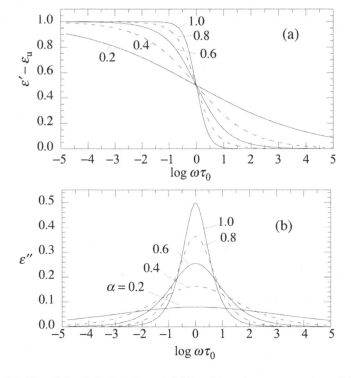

Figure 2.9 The Cole–Cole function: real (a) and imaginary components (b) of the relative permittivity plotted against log(circular) frequency for several values of the broadening parameter.

and

$$\varepsilon'' = (\varepsilon_{\mathrm{r}} - \varepsilon_{\mathrm{u}}) \left(\frac{(\omega\tau_0)^\alpha \sin\left(\frac{\pi}{2}\alpha\right)}{1 + 2(\omega\tau_0)^\alpha \cos\left(\frac{\pi}{2}\alpha\right) + (\omega\tau_0)^{2\alpha}} \right). \qquad (2.40)$$

Both ε' and ε'' are plotted in Figure 2.9 for several values of the broadening parameter.

It is often useful to make a *complex plane* plot of ε' and ε'' where a value of the imaginary component, $\varepsilon''(\omega)$, is plotted against the real part $\varepsilon'(\omega)$ at the same frequency, ω. Such complex plane plots are often called *Argand* diagrams. Complex plane plots for the Cole–Cole function are shown in Figure 2.10. It can be demonstrated that for $\alpha = 1$ the plot is a semi-circle (when plotted on equal x-axis scales as in Figure 2.10). For $\alpha < 1$ the plots are circular arcs whose intercepts on the ε'-axis are ε_{r} and ε_{u}. The heights above the ε'-axis become progressively smaller compared to $\varepsilon_{\mathrm{r}} - \varepsilon_{\mathrm{u}}$ as α decreases. An $\varepsilon''(\omega)$ vs. $\varepsilon'(\omega)$ complex plane plot for the

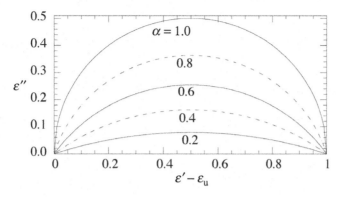

Figure 2.10 Complex plane plot of the Cole–Cole function for various values of the broadening parameter.

Cole–Cole function, eq. (2.38) is commonly known as a "Cole–Cole" plot. Such complex plane plots of experimental dielectric data, whether or not the Cole–Cole function itself is invoked, have also become commonly known as "Cole–Cole" plots.

It is to be noted in Figure 2.9 that the loss peak ε'' is symmetric about $\omega\tau_0 = 1$ on a log ω plot. However, it is characteristic of the glass transition region in amorphous polymers that the dielectric loss peak is actually skewed toward high frequency. Davidson and Cole showed that a modification of the SRT function introduces this type of asymmetry [10]. The skewing parameter, β, is an exponent that encompasses the entire denominator of the SRT function, or

$$\text{Davidson–Cole} \quad \varepsilon^*(i\omega) = \varepsilon_u + (\varepsilon_r - \varepsilon_u)/[1 + i\omega\tau_0]^\beta. \tag{2.41}$$

The Havriliak and Negami (abbreviated often in this work as "HN") equation incorporates both exponents into a single equation [11]. Thus,

$$\text{Havriliak–Negami} \quad \varepsilon^*(i\omega) = \varepsilon_u + (\varepsilon_r - \varepsilon_u)/[1 + (i\omega\tau_0)^\alpha]^\beta, \tag{2.42}$$

which leads to

$$\varepsilon' - \varepsilon_u = (\varepsilon_r - \varepsilon_r)Z^{-\beta}\cos\beta\Theta, \tag{2.43}$$

$$\varepsilon'' = (\varepsilon_r - \varepsilon_u)Z^{-\beta}\sin\beta\Theta, \tag{2.44}$$

where

$$\Theta = \tan^{-1}\left(\frac{(\omega\tau_0)^\alpha \sin\dfrac{\pi}{2}\alpha}{1 + (\omega\tau_0)^\alpha \cos\dfrac{\pi}{2}\alpha}\right) \tag{2.45}$$

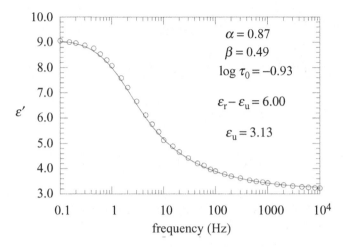

Figure 2.11   The PVAc data in Figure 2.8 transformed to the frequency domain: real component. The curves are HN function fits to the data with the parameters shown. The latter were determined via numerical regression.

and

$$Z = \left[1 + 2(\omega\tau_0)^\alpha \cos\left(\frac{\pi}{2}\alpha\right) + (\omega\tau_0)^{2\alpha} \right]^{1/2}. \tag{2.46}$$

Values of $\beta < 1$ lead to skewing of the loss component toward high frequency and values of $\beta > 1$ give skewing in loss toward low frequency. It is found that both the high frequency skewing brought about by $\beta < 1$ and the symmetric broadening associated with $\alpha < 1$ are necessary to fit frequency domain dielectric data in the glass transition region in amorphous polymers. Thus the HN function has been much used in this context. As an example, the data in Figure 2.8, when Fourier transformed, lead to the data points in Figure 2.11, Figure 2.12, and Figure 2.13. It is seen that the HN function, with the proper choice of parameters, leads to an excellent fit of the data in Figure 2.11, Figure 2.12, and Figure 2.13, including the high frequency "tail" in the loss and the asymmetry of the complex plane plot. The KWW function in order to fit the time domain data as in Figure 2.8 must inherently incorporate both the symmetric broadening and the skewing but with one parameter, β rather than the two, i.e., α and β, in the HN function. Systematic discrepancies in the KWW fit may be seen in Figure 2.8 that are not present in the HN fits in Figure 2.11 and Figure 2.13. Thus it is concluded that the KWW function does remarkably well in introducing the skewing and broadening but, with another parameter, the HN function does noticeably better in fitting the data.

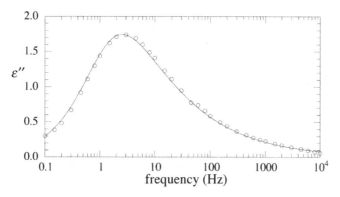

Figure 2.12 The PVAc data in Figure 2.8 transformed to the frequency domain: loss component. The curves are HN function fits to the data with the same parameters as in Figure 2.11.

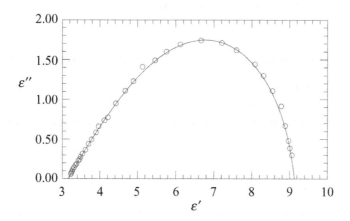

Figure 2.13 Complex plane plot of the data in Figures 2.11 and 2.12 with the HN function fit.

Another empirical function was proposed by Fuoss and Kirkwood [12] based on modification of the loss component of the Debye SRT equation, eq. (2.34), where the $\omega\tau$ terms are replaced by $(\omega\tau_0)^m$. The result can be expressed as

$$\text{Fuoss–Kirkwood} \quad \varepsilon''(\omega) = \varepsilon''_{\max}\,\text{sech}(m(\ln\omega\tau_0)). \tag{2.47}$$

However, there appears to be no closed form expression for the real component $\varepsilon'(\omega)$. This severely limits the utility of the equation. The Fuoss–Kirkwood equation has been modified by Jonscher [13] to contain two exponent parameters and thus

3

NMR spectroscopy

Nuclear magnetic resonance (NMR) is the primary spectroscopic technique utilized in the study of polymer dynamics. NMR is in many ways complementary to the scattering (neutron) and relaxation (mechanical and dielectric) techniques described in Chapters 1 and 2. The major advantage of NMR over other methods of characterizing polymer dynamics is its selectivity. Firstly, the precession (Larmor) frequencies of different nuclei, and even different isotopes of the same element, differ dramatically. Secondly, the resonance of a given nucleus depends upon its surroundings due to internal coupling, making NMR sensitive to the details of the chemical structure of the polymer. Finally, the dramatic differences in the natural abundances of different isotopes provide the opportunity to increase selectivity through isotopic labeling.

NMR provides information on the dynamics of local motions. Many NMR parameters are sensitive to local molecular motions and the NMR methods that have been applied in probing polymer dynamics are varied and numerous [1,2]. NMR parameters that are sensitive to local molecular motions include relaxation times, spectrum line shape, the strength of dipolar coupling, and chemical-shift anisotropy. The accessible spectral windows depend upon the type of measurement performed, ranging from 10^{-1} Hz for measurements sensitive to slow motions to several hundred megahertz to those sensitive to fast motions. In the case of polymers at temperatures well above the glass transition temperature local motions are fast processes that average chemical-shift anisotropy, homonuclear and heteronuclear dipolar couplings, and quadrupolar couplings. Therefore, conventional ^{13}C spectroscopic techniques applied to solutions that probe modes in the Larmor frequency region, namely spin-lattice relaxation time and nuclear Overhauser-effect (NOE) measurements, are applicable. For the much slower motions that occur nearer the glass transition, or for subglass relaxations, it is the slow averaging, or partial averaging, of anisotropic interactions that is of primary interest.

Because of complex couplings, particularly intermolecular couplings, the dynamics probed by many NMR measurements are difficult to interpret molecularly.

In this book we will concentrate on NMR methods that are well suited for molecular interpretation, thereby providing valuable mechanistic information about polymer relaxation processes that cannot be provided through other experimental means. These methods are also often well suited for direct comparison with molecular simulations, and synergism between NMR measurements and molecular simulations can provide important mechanistic insight into relaxation processes in polymers. The techniques covered in this chapter are ^{13}C spin-lattice and NOE measurements and exchange NMR with emphasis on 2D exchange. The relaxation methods probe relaxation time scales from nanoseconds to microseconds, while the 2D methods primarily cover relaxation time scales from milliseconds to tens of seconds. ^{2}H line shape analysis [1] and ^{1}H double quantum spectroscopy [3], which also provide important information about polymer dynamics, are not covered because the information provided by these methods can be obtained largely by the relaxation and exchange methods and because they have been applied primarily to dynamics associated with topological constraints (reptation dynamics) that are not covered in this book.

3.1 NMR basics

3.1.1 Influence of applied magnetic fields on nuclear magnetic moments

When a nucleus of magnetic moment μ is placed in a static magnetic field of strength $\mathbf{B}_0$, the magnetic moment of the nucleus becomes directionally quantized. The component of the magnetic moment in the field direction (z-axis) is given by

$$\mu_z = m\gamma\hbar, \tag{3.1}$$

where $m = I, I - 1, \ldots, -I$ is the directional quantum number. The angular moment quantum number (I) and gyromagnetic ratio (γ) for the important nuclides considered in this book are given in Table 3.1 The nuclear dipoles precess (in the classical representation) around the field direction with a precession (Larmor) frequency given by

$$\omega_{\mathrm{L}} = |\gamma|\, B_0. \tag{3.2}$$

Directional quantization restricts the precessional angle to specific values (e.g., 54.44° for $I = \frac{1}{2}$ nuclei).

The application of the static field B_0 results in a difference in energy levels of a nucleus depending upon its directional quantum number. The energy between two adjacent energy levels is

$$\Delta E = \gamma\hbar B_0. \tag{3.3}$$

Table 3.1. *Magnetic properties of important NMR sensitive nuclides*

			Larmor frequency (MHz)		
Nuclide	Spin (I)	$\gamma(10^7 \text{ rad T}^{-1} \text{ s}^{-1})$	$B_0 = 2.35$ T	$B_0 = 11.75$ T	$B_0 = 18.80$ T
^{1}H	½	26.7519	100	500	800
^{2}H	1	4.1066	15.35	76.75	122.8
^{12}C	0				
^{13}C	½	6.7283	25.14	125.7	201.1

Data from Friebolin [4].

When an electromagnetic field of magnitude B_1 and frequency ω_1 is superimposed on the static field, the photon energy ($\hbar\omega_1$) engenders transitions of the nuclear dipoles from lower to higher and higher to lower (adjacent) energy levels (with equal probability) when the resonance condition, $\Delta E = \hbar\omega_1$, or $\omega_1 = \omega_L$, is achieved. Since the static field $\mathbf{B}_0$ results in an excess of spins in the low energy state (and hence a macroscopic magnetization in the direction of $\mathbf{B}_0$) the net result of the field $\mathbf{B}_1$ is an adsorption of photons.

3.1.2 Chemical shift anisotropy and dipolar coupling

In molecules nuclei are magnetically shielded by their neighbors, hence the effective field at the position of a nucleus is weaker than the externally applied field. This shielding results in a chemical shift of a nuclear spin (change in resonance frequency) and the site-selectivity of NMR spectra. The tensorial nature of the shielding interactions, and hence chemical shift, results in a dependence of the chemical shift upon the orientation of the molecular segment to which the nucleus belongs (and to which the interaction tensors are generally fixed) with respect to the external magnetic field. In a liquid, due to rapid molecular motions (relaxation), this tensorial anisotropy is averaged to zero and the observed chemical shift is the trace of the chemical-shift tensor. In powder (solid-state) NMR all orientations have the same probability due to the absence of relaxation and the signal obtained for each nuclei is the sum of the elementary chemical shifts corresponding to the different orientations. The angular dependence of the NMR frequency for a particular nucleus in high magnetic fields is given by [1]

$$\omega = \omega_L + \Delta\omega, \tag{3.4}$$

$$\Delta\omega(\theta, \phi) = \omega_d\left[\tfrac{1}{2}(3\cos^2\theta - 1 - \eta\sin^2\cos 2\phi)\right]. \tag{3.5}$$

Here the Larmor frequency ω_L includes isotropic chemical shifts, ω_d is the strength of the isotropic coupling and $\Delta\omega$ is the orientation dependent coupling which

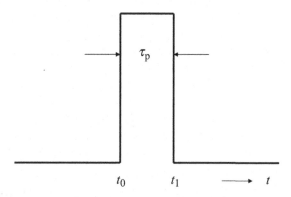

Figure 3.1 Schematic of an NMR pulse. The RF generator is switched on at time t_0 for a duration τ_p typically on the order of microseconds.

averages to zero in liquids. The asymmetry parameter η describes the deviation of the anisotropic coupling from axial symmetry. The angles θ, ϕ are the polar angles of the magnetic field $\mathbf{B}_0$ in the principal axes system of the coupling tensor. This in turn is often related in a simple way to the molecular geometry, e.g., the unique axis (z) may lie along a C–H bond. In solids a similar averaging effect to that manifested in liquids can be achieved by mechanically rotating the sample at the "magic" angle of $\theta_M = 54.74°$ relative to $\mathbf{B}_0$ for which the angular function $3\cos^2\theta - 1$ vanishes. Such magic angle spinning (MAS) yields liquid-like spectra if the spinning frequency ω_R is significantly larger than the width of the anisotropic powder pattern ($\omega_R \gg \omega_d$).

3.2 The pulsed NMR method

3.2.1 Application of pulses

In practice NMR experiments utilize a series of radiofrequency (RF) pulses of the appropriate frequency. A schematic representation of an RF pulse is shown in Figure 3.1. These pulses excite all nuclei of one species simultaneously. The pulse is generated at frequency ω_1 and has a width in the frequency domain approximately proportional to $1/\tau_p$. To induce transitions, the RF pulse is applied to the sample perpendicular to the static field (say along the x-direction). The magnetic vector of the RF pulse ($\mathbf{B}_1$) interacts with the nuclear dipoles, and hence the macroscopic magnetization, $\mathbf{M}_0$, given as the sum of the nuclear dipoles, which is initially in the direction of the static field $\mathbf{B}_0$ (z-direction). Under the influence of $\mathbf{B}_1$, $\mathbf{B}_0$ is tipped away from the z-axis. In the rotating coordinate system x', y', and z, which rotates with frequency ω_1 about the z-axis, the effect of $\mathbf{B}_1$ (which lies along the x'-axis),

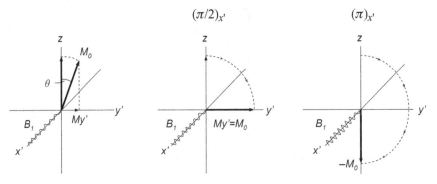

Figure 3.2 Rotation of the macroscopic magnetization under $(\pi/2)_{x'}$ and $(\pi)_{x'}$ pulses. The field $\mathbf{B}_1$ is applied along the x'-axis. Adapted from Friebolin [4].

is to rotate $\mathbf{B}_0$ in the y',z-plane. The angle of rotation (pulse angle) is given by the relationship

$$\theta = \gamma B_{1i} \tau_{\mathrm{p}}, \tag{3.6}$$

where B_{1i} is the pulse amplitude at the resonance frequency (ω_i) of nuclei type i. Most pulse techniques utilize combinations of $90°$ and/or $180°$ pulses. A $90°$ pulse along the x'-axis, $(\pi/2)_{x'}$, rotates $\mathbf{M}_0$ from the z-axis to the y'-axis, resulting in a maximum in the transverse component of the magnetization $(M_{y'} = M_0)$, as illustrated in Figure 3.2. Since detection of the NMR signal is proportional to the transverse magnetization, this pulse results in a signal maximum. A $180°$ pulse along the x'-axis $(\pi_{x'})$ reverses $\mathbf{M}_0$ from the positive z-axis to the negative z-axis, resulting in no detectable signal.

3.2.2 Relaxation of the macroscopic magnetization

At the end of a pulse the macroscopic magnetization vector $\mathbf{M}_0$ is at an angle θ from its equilibrium direction along the z-axis as given by eq. (3.6). When the perturbing $(\mathbf{M}_1)$ pulse is turned off, the macroscopic magnetization (spin system) relaxes back to its original state according to the Bloch equations [4]:

$$\frac{\mathrm{d}M_z}{\mathrm{d}t} = -\frac{M_z - M_0}{T_1}[M_z(t \to \infty) = M_0], \tag{3.7}$$

$$\frac{\mathrm{d}M_{x'}}{\mathrm{d}t} = -\frac{M_{x'}}{T_2}[M_{x'}(t \to \infty) = 0], \tag{3.8}$$

$$\frac{\mathrm{d}M_{y'}}{\mathrm{d}t} = -\frac{M_{y'}}{T_2}[M_{y'}(t \to \infty) = 0], \tag{3.9}$$

where $M_{x'}$ and $M_{y'}$ are the transverse components of the macroscopic magnetization in the rotating coordinate system. T_1 is the spin-lattice or longitudinal relaxation time whose measurement plays a major role in probing polymer dynamics, as discussed in Sec. 3.3. T_2 is the spin–spin or transverse relaxation time, which controls the rate of decay of phase coherence and the measured NMR signal.

3.2.3 Free induction decay (FID)

Immediately after a $(\pi/2)_{x'}$ pulse the z-component of the macroscopic magneti-zation is zero due to an equalization of spin populations. Instead of M_z there is a transverse magnetization $M_{y'}$, which is the measured NMR signal. The nuclear dipoles exhibit phase coherence immediately after the pulse. The phase coherence (and hence the net transverse magnetization) is lost according to eq. (3.9). The energy of the system is not altered by spin–spin relaxation since the spin popula-tions are not affected. The main contribution to spin–spin relaxation comes from heterogeneities in the local magnetic field causing chemically equivalent nuclei to precess at different rates (some faster, some slower than the mean Larmor fre-quency), resulting in loss of phase coherence.

Equation (3.9) indicates that the measured NMR signal should exhibit simple first order decay. However, the real signal will show maxima with a spacing (in time) that is the inverse of the difference between the frequency of the applied pulse (ω_1) and the resonance frequency of the nuclei (ω_i), i.e., $\Delta t = 2\pi \Delta \omega$, as illustrated in Figure 3.3. The measured decay of the transverse magnetization is referred to as free induction decay (FID). For samples containing nuclei with different resonance frequencies due to different environments or spin–spin coupling, the FID spectra can become quite complicated. However, the real component of the Fourier transform of the FID spectra [4],

$$g(\Delta\omega) = \text{Re} \left[\int_0^\infty FID(t) \exp[-i\Delta\omega]\, dt \right], \qquad (3.10)$$

yields the familiar frequency domain spectrum, as shown in Figure 3.3 for the case of a single resonance frequency.

3.3 NMR relaxation measurements

3.3.1 Spin-lattice relaxation times

Spin-lattice relaxation times for ^{13}C nuclei provide a particularly useful and selec-tive probe of polymer dynamics. This is because: (a) the ^{13}C resonances differ sig-nificantly for different bonding situations, allowing probing of different, chemically

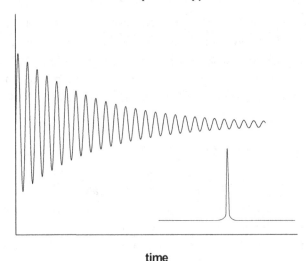

time

Figure 3.3 FID for a material with a single resonance frequency and its associated Fourier transform.

inequivalent locations on the polymer chain; (b) T_1 values can vary from milliseconds to hundreds of seconds depending upon local polymer dynamics; and (c) the spectra and the interpretation of the relaxation mechanism are relatively simple. The interpretation of ^{13}C spin lattice relaxation is greatly simplified due to the dominance of the dipole–dipole (dipolar) relaxation mechanism. The motion of surrounding magnetic dipoles, dominated by protons attached to the ^{13}C nuclei, results in fluctuations in the magnetic field at the position of each ^{13}C nuclei. The components of this fluctuation at the appropriate frequency (resonance frequency of the ^{13}C nuclei, see Table 3.1) can induce nuclear spin transitions and hence help reestablish equilibrium in the spin populations and decay of the longitudinal magnetization, as described by eq. (3.7). The faster the motion of the attached protons, the less "spectral power" at the ^{13}C resonance frequency, and the longer the spin-lattice relaxation time T_1. While the time scale for the experimentally measured decay of the magnetization of a ^{13}C nucleus in a polymer melt is typically on the order of seconds, the corresponding decay of the $^{13}C–^1H$ vector autocorrelation function is on the order of nanoseconds.

The spin-lattice relaxation time is related to the reorientation of $^{13}C–^1H$ vectors through the relationship [5]

$$\frac{1}{nT_1} = K[J(\omega_H - \omega_C) + 3J(\omega_C) + 6J(\omega_H + \omega_C)]. \qquad (3.11)$$

Here n is the number of attached protons, ω_H and ω_C are the proton and ^{13}C resonance angular frequencies, given as 2π times the values given in Table 3.1.

The constant K assumes the value of 2.29×10^9 and 2.42×10^9 s^{-2} for sp^3 and sp^2 nuclei, respectively. For ^{2}H nuclei, spin-lattice relaxation is dominated by electric quadrupole coupling. The relationship between the spin relaxation time and the reorientation of the C–^{2}H bonds is given as

$$\frac{1}{T_1} = \frac{3}{10} \pi^2 K_Q^2 [J(\omega_D) + 4J(2\omega_D)]. \tag{3.12}$$

The quadrupole coupling constant is taken as 172 kHz for ^{2}H bonded to sp^3 and 190 kHz for ^{2}H bonded to sp^2 carbon nuclei. Here, ω_D is the deuteron resonance (angular) frequency, given as 2π times the values given in Table 3.1. The spectral density $J(\omega)$ as a function of angular frequency ω is given by

$$J(\omega) = \tfrac{1}{2} \int\limits_{-\infty}^{\infty} P_2(t) e^{i\omega t} \, dt \tag{3.13}$$

and the orientational autocorrelation function $P_2(t)$ is

$$P_2(t) = \tfrac{1}{2} \{ 3 \langle [\mathbf{e}_{CH}(t) \cdot \mathbf{e}_{CH}(0)]^2 \rangle - 1 \}, \tag{3.14}$$

where $\mathbf{e}_{CH}(t)$ is the unit vector along a ^{13}C–^{1}H or C–^{2}H bond at time t.

The spin-lattice relaxation time is most often measured using the inversion recovery method [4]. The pulse sequence used is $(mT_1 - (\pi)'_x - \tau - (\pi/2)'_x - \text{FID})_n$, illustrated in Figure 3.4. The subscript n indicates that multiple spectra, separated by many multiples of $T_1(mT_1)$ are collected and averaged. The $(\pi)_{x'}$ pulse causes M_0 to lie in the $-z$ direction ($M_z = -M_0$). During the waiting time τ the system relaxes according to eq. (3.7). A signal is induced in the receiver not by M_z but by the transverse magnetization. Hence, the second pulse $(\pi/2)_{x'}$ rotates $M_z(\tau)$ into $M_{y'}$ so that it can be detected. Normally five–ten values of τ are utilized. The relative signals for the various delay times are shown in Figure 3.4. From the magnitude of the signal as a function of waiting time, i.e., $M_z(\tau)$, application of eq. (3.7) allows the determination of the spin-lattice relaxation time T_1.

3.3.2 Nuclear Overhauser effect (NOE)

^{13}C NMR spectra are normally recorded with ^{1}H broadband (BB) decoupling, as illustrated in Figure 3.4. In practice ^{13}C$\{^1$H$\}$ decoupling is performed by irradiating with a continuous sequence of "composite pulses" that saturate the protons, i.e., result in equal populations of both proton energy levels. The net result is a much simplified spectrum due to elimination of ^{13}C$\{^1$H$\}$ coupling, which divides already low intensity ^{13}C signals (due to the low natural abundance of ^{13}C) into multiplets.

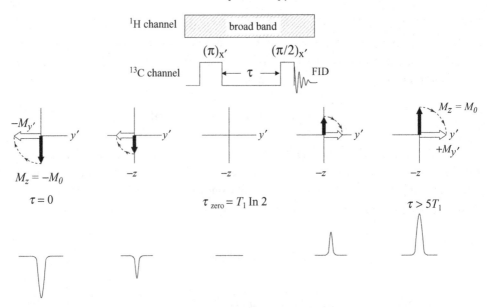

Figure 3.4 Pulse sequence for the T_1 inversion recovery, magnetization vector diagram and the detectable signal for various τ values from the Fourier transform of the FID. Adapted from Friebolin [4].

Broadband (BB) decoupling also yields NOE amplification of the signals. In the case of pure dipolar relaxation the NOE (signal amplification) is given as [5]

$$NOE = 1 + \frac{\gamma_{^1H}}{\gamma_{^{13}C}} \frac{6J(\omega_H + \omega_C) - J(\omega_H - \omega_C)}{J(\omega_H - \omega_C) + 3J(\omega_C) + 6J(\omega_H + \omega_C)}. \qquad (3.15)$$

Hence the extent of enhancement gives additional information about the dynamics. In the limit of rapid reorientation of the molecule (in particular, reorientation of $^{13}C-^1H$ bond vectors) compared to the resonance frequencies of nuclei, the NOE is given as

$$NOE = 1 + \frac{1}{2}\frac{\gamma_{^1H}}{\gamma_{^{13}C}} = 2.99. \qquad (3.16)$$

Qualitatively, saturation of the proton transitions results in increased populations of $^{13}C\{^1H\}$ spin pairs that can relax through double quantum transitions, i.e. with both the ^{13}C nucleus and the attached proton in the same spin state flipping simultaneously, or ($^{13}C\{^1H\}\beta\beta \rightarrow \alpha\alpha$), or through zero quantum transitions ($\alpha\beta \rightarrow \beta\alpha$). The former transitions depend upon the spectra power of magnetic fluctuations (and hence local dynamics) at frequency $\omega_C + \omega_H$, and increase the signal strength at ω_C, while the latter transitions depend upon the spectra power of magnetic fluctuations at frequency $\omega_H - \omega_C$ and decrease the signal strength at ω_C. The NOE itself can be

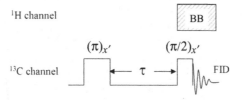

Figure 3.5 Pulse sequence for an NOE measurement.

measured by comparing the signal intensities obtained with continuous ^{1}H broad band decoupling with those obtained using inverse gated decoupling, as illustrated in Figure 3.5. Here, the ^{1}H BB decoupler is utilized only during the observing pulse and subsequent recording of the FID of the ^{13}C signal. As a consequence, ^{13}C$\{^1$H$\}$ coupling is eliminated but no NOE is allowed to build up.

3.3.3 Example application of NMR relaxation measurements

One of the strengths of NMR is the ability to resolve different chemical environments. This chemical specificity is clearly demonstrated in NMR spin-lattice relaxation studies of 1,4-poly(butadiene) (PBD). The dynamics and relaxational behavior of PBD have been the subject of extensive experimental study with a variety of techniques, including NMR, dielectric relaxation, and dynamic neutron scattering. PBD is a good glass former, and its simple chemical structure, narrow molecular weight distribution, and wide variety of available microstructures make it ideal for investigations of the glass transition as well as subglass and high temperature dynamics. The various resolvable NMR resonances in PBD are shown in Figure 3.6.

A comparison of T_1 values for PBD from experiment and molecular dynamics (MD) simulation (see Sec. 6.4 for additional details on MD simulations of PBD) has been made over the temperature range 273–353 K. Results for the highest and lowest temperatures are shown in Figure 3.6. Excellent agreement is seen for all resolvable resonances over the entire temperature range. The agreement between experiment and simulation for T_1 for all resolvable resonances indicates that conformational dynamics, which are primarily responsible for the reorientation of the C–H vectors, are well reproduced by the simulations over the entire temperature range investigated for each type of dihedral in PBD, i.e. $C_{sp^3} - C_{sp^3} \Leftrightarrow C_{sp^2} \overset{trans}{=} C_{sp^2}$ (*trans* allyl), $C_{sp^3} - C_{sp^3} \Leftrightarrow C_{sp^2} \overset{cis}{=} C_{sp^2}$ (*cis* allyl) and $C_{sp^2} - C_{sp^3} \Leftrightarrow C_{sp^3} - C_{sp^2}$ (alkyl) dihedrals. This detailed, quantitative agreement between simulation and experiment is critical to the validation of simulations, the interpretation of experiments, and the assignment of relaxation mechanisms. The value of coordinated experimental and simulation studies of polymer relaxations is considered in detail in Chapters 6 and 8.

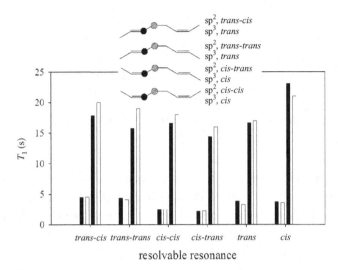

Figure 3.6 ^{13}C spin-lattice relaxation times in 1,4-PBD ($M_w = 1600$) for resolvable resonances. Long times were measured at 353 K, short times at 273 K. Solid bars are times obtained from MD simulations. The chemical environment for each resolvable NMR resonance for sp^2 (black circles) and sp^3 (grey circles) ^{13}C nuclei is shown. Data from Smith *et al.* [6].

3.4 NMR exchange spectroscopy

3.4.1 2D exchange measurements

NMR relaxation measurements are sensitive to motions whose frequencies are in the regime of the resonance frequencies of the relaxing nuclei, corresponding to polymer relaxations well above T_g. At lower temperatures line broadening and short T_1 times make spectra very difficult to measure and resolve. Techniques for measuring relatively slow subglass motion and relaxations near the glass transition temperature take advantage of the dependence of NMR resonance frequencies on the strength of dipolar coupling and chemical-shift anisotropy. Due to the dependence of the NMR frequency of a nucleus on the orientation of a given molecular segment relative to $\mathbf{B}_0$ (eq. (3.5)), a change in molecular orientation in general causes a frequency change that can be detected in 2D exchange experiments [1,7,8]. It is crucial to realize that the anisotropic NMR frequencies do not reflect the orientations of the nuclear spins, but rather the orientations of molecular segments to which the interaction tensors are fixed. The exchange experiment correlates a coupling tensor (and hence a particular NMR frequency) at the beginning of a "mixing period" with the same tensor at the end of the period. Hence, the experiment correlates the orientations of the *same* tensor at *different* points in time.

The pulse sequence for a 2D NMR exchange experiment is shown in Figure 3.7. The preparation and data acquisition phases are separated by an evolution and

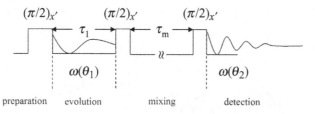

Figure 3.7 Pulse sequence for a 2D exchange measurement. Adapted from Friebolin [4].

mixing phase. The development of the nuclear spin system occurs during the evolution period τ_1 at the beginning of the pulse sequence. Incrementing the evolution period τ_1 provides the basis for the first frequency (chemical-shift) dimension $\omega_1(\theta_1)$ through a Fourier transform. The NMR signal is detected in the period τ_2 at the end of the pulse sequence, providing the basis for the second frequency dimension $\omega_2(\theta_2)$. A variable mixing period of duration τ_m, during which dynamic processes (relaxations) can take place, is inserted between evolution and detection. Varying τ_m allows slow dynamic processes with a duration in the range of milliseconds–seconds to be followed.

The 2D spectrum $S(\omega(\theta_1), \omega(\theta_2); \tau_m)$ represents the probability of finding a nucleus with a frequency ω_1, determined by the local segmental orientation θ_1 (via eq. (3.5)), having, after a time interval τ_m, a frequency ω_2 determined by a (perhaps new) orientation θ_2. The frequency measurement typically requires evolution and detection times of the order of a millisecond, while the mixing time can be from milliseconds to tens of seconds. 2D exchange measurements are typically done on ^{2}H since aliphatic C–^{2}H bonds result in axially symmetric coupling ($\eta = 0$ in eq. (3.5)). The asymmetry parameter η for the ^{13}C anisotopic chemical shift can also be small, depending upon the local chemical environment. Axially symmetric or near axially symmetric coupling greatly simplifies the interpretation and modeling of the 2D spectrum. The angular reorientation distribution provided by the 2D exchange measurements can then be interpreted molecularly in order to obtain insight into polymer segmental dynamics.

3.4.2 Example applications of 2D NMR to study slow relaxations in polymers

Modeling of 2D exchange spectra is typically carried out by assuming isotropic diffusion with a distribution of relaxation times. Angular changes ($\Delta\theta$) can be assumed to occur as a consequence of discrete jumps and/or a distribution of angular excursions. For these cases, modeling of the 2D spectra consists of reproducing the spectra via eq. (3.5) with a distribution of angular reorientations $R[\Delta\theta(\tau_m)]$, where $0 \leq \Delta\theta(\tau_m) \leq \pi/2$, and an associated distribution of reorientation times.

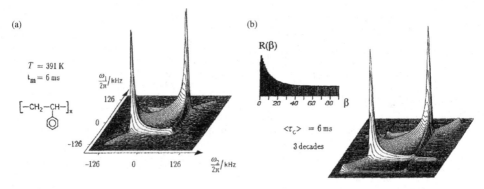

Figure 3.8 2D exchange spectra of PS at 391 K: (a) experiment; (b) simulation. Also shown is the distribution of angular jumps used to model the experimental spectra and the associated model spectra. Reproduced with permission from K. Schmidt-Rohr and H. W. Spiess [1].

This approach has been applied to 2D ^{2}H spectra for deuterated poly(styrene) (PS) as a function of temperature approaching the glass transition temperature. The measured and modeled 2D spectra at 391 K for a mixing time of 6 ms are shown in Figure 3.8. When most of the 2D exchange intensity lies along the diagonal, $\Delta\theta(\tau_m)$ is small, implying little reorientation has occurred, while off-diagonal intensity indicates reorientation. The experimental spectrum is well reproduced with the distribution of angular reorientations for the C–^{2}H vectors shown in Figure 3.8 and reorientation time spectra covering three decades with a mean time of 6 ms.

References

[1] K. Schmidt-Rohr and H. W. Spiess, *Multidimensional Solid-State NMR and Polymers* (New York: Academic Press, 1994).
[2] H. W. Spiess, *Macromol. Chem. Phys.*, **204**, 340 (2003).
[3] H. W. Spiess, *Encyclopedia of Nuclear Magnetic Resonance*, Vol. 9 (New York: Wiley, 2002).
[4] H. Friebolin, *Basic One- and Two-Dimensional NMR Spectroscopy* (New York: Wiley-VCH, 1998).
[5] D. J. Gisser, S. Gowinkowski, and M. D. Ediger, *Macromolecules*, **24**, 4270 (1991).
[6] G. D. Smith, O. Borodin, D. Bedrov, W. Paul, X. Qiu, and M. D. Ediger, *Macromolecules*, **34**, 5192 (2001).
[7] H. W. Spiess, *Annu. Rep. NMR Spect.*, **34**, 1 (1997).
[8] K. Schmidt-Rohr and H. W. Spiess, *Annu. Rep. NMR Spectr.*, **48**, 1 (2002).

4

Dynamic neutron scattering

Dynamic neutron scattering is distinguished from other experimental probes of polymer motion by its spatial sensitivity and its ability to probe dynamics on very short time scales. By varying momentum transfer, dynamic neutron scattering can be used to probe the time evolution of atomic motions on length scales ranging from that of the monomer to the radius of gyration of the polymer and beyond. The time scale accessible to dynamic neutron scattering has been extended significantly, with neutron scattering methods now covering time scales from picoseconds to hundreds of nanoseconds. Furthermore, dynamic neutron scattering is sensitive to isotopic substitution, making hydrogen/deuterium exchange a powerful tool for probing the motion of a particular component of a polymer or of a single polymer chain.

4.1 Neutron scattering basics

4.1.1 Coherent and incoherent elastic scattering

A neutron with velocity v has an associated wave vector $\mathbf{k}$ with amplitude

$$k = \frac{2\pi}{\lambda},$$

(4.1)

yielding a momentum and kinetic energy

$$\mathbf{p} = \hbar\mathbf{k},$$

$$E = \frac{1}{2}mv^2 = \frac{\hbar^2 k^2}{2m} = k_B T = \hbar\omega,$$

(4.2)

respectively, where the de Broglie wavelength is given by

$$\lambda = \frac{h}{mv}.$$

(4.3)

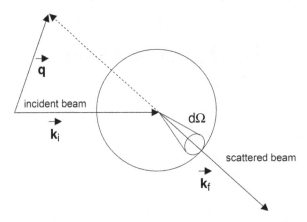

Figure 4.1 The geometry of a scattered neutron beam. Adapted from Higgins and Benoît [1].

Here m is the neutron mass and v its velocity. The temperature of thermal neutrons depends on the source and is typically around 300 K, yielding a neutron wavelength of 2 Å, which can be varied between about 0.4 Å and 30 Å. The amplitude of a beam of neutrons at point $\mathbf{r}$ in space and t in time is

$$\psi_i(\mathbf{r}, t) = \exp i(\omega t - \mathbf{r} \cdot \mathbf{k}_i). \tag{4.4}$$

A single nucleus scatters the incident neutrons into a spherical wave (centered on the nucleus) given by

$$\psi_f(\mathbf{r}, t) = \frac{b}{r} \exp i(\omega t - r k_f) \tag{4.5}$$

since the vectors $\mathbf{k}_f$ and $\mathbf{r}$ have the same direction (see Figure 4.1). The quantity b is called the scattering length and depends upon the nucleus (atomic number and isotope). The square of the amplitude of the scattered wave at a distance r from the nuclei is given as b^2/r^2. Hence, the probability that a neutron is scattered into solid angle $d\Omega$, shown in Figure 4.1, is b^2.

For a collection of N nuclei, each having scattering length b_k, the scattered wave is described by

$$\psi_f(\mathbf{r}) = \sum_{k=1}^{N} \frac{b_k}{r_k} \exp i(\mathbf{q} \cdot \mathbf{r}_k), \tag{4.6}$$

where r_k is the vector from the observer (at position $\mathbf{r}$ relative to the sample) and $\mathbf{q}$ is the scattering vector (momentum transfer) defined in Figure 4.1. The differential scattering cross-section (number of neutrons scattered into solid angle $d\Omega$ divided by the incident neutron flux) is given by [1]

$$
\frac{d\sigma}{d\Omega} = \left\langle \sum_k^N b_k \exp i(\mathbf{q} \cdot \mathbf{r}_k) \sum_j^N \exp i(-(\mathbf{q} \cdot \mathbf{r}_k)) \right\rangle
$$

$$
= \left\langle \sum_{j,k}^N b_j b_k \exp i(\mathbf{q} \cdot (\mathbf{r}_k - \mathbf{r}_j)) \right\rangle,
\tag{4.7}
$$

where the brackets indicate an average over all configurations of the sample during the time of the experiment and r_k has been replaced by r as the distance to the observer is much greater than the sample size. The double sum contains two parts, namely a self-scattering term ($j = k$) and a distinct term ($j \neq k$) [1]

$$
\frac{d\sigma}{d\Omega} = \sum_k^N b_k^2 + \left\langle \sum_{j\neq k}^N b_j b_k \exp i(\mathbf{q} \cdot (\mathbf{r}_k - \mathbf{r}_j)) \right\rangle.
\tag{4.8}
$$

In materials where the scattering is dominated by one element, the different scattering lengths associated with different isotopes (chemical or spin) of that element are independent of atomic positions. In this case the elastic differential scattering cross-section is

$$
\frac{d\sigma}{d\Omega} = N\langle b^2 \rangle + \langle b \rangle^2 \left\langle \sum_{j\neq k}^N \exp i(\mathbf{q} \cdot (\mathbf{r}_k - \mathbf{r}_j)) \right\rangle,
\tag{4.9}
$$

where

$$
\langle b^2 \rangle = \frac{1}{N} \sum_i^N b_i^2,
\tag{4.10}
$$

$$
\langle b \rangle = \frac{1}{N} \sum_i^N b_i.
\tag{4.11}
$$

Defining

$$
\Delta b^2 = \langle b^2 \rangle - \langle b \rangle^2,
\tag{4.12}
$$

eq. (4.9) can be rewritten as [1]

$$
\frac{d\sigma}{d\Omega} = N\Delta b^2 + \langle b \rangle^2 \left\langle \sum_{j,k}^N \exp i(\mathbf{q} \cdot (\mathbf{r}_k - \mathbf{r}_j)) \right\rangle.
\tag{4.13}
$$

The first term corresponds to incoherent scattering and requires the presence of nuclei with different scattering lengths. An isotopically pure sample containing nuclei without spin would show no incoherent scattering. The second term is the coherent scattering, which depends upon the structure of the sample (positions occupied by the various nuclei) and yields an angular dependence of the scattering.

Note that when different elements (e.g., carbon and hydrogen) contribute significantly to the scattering, or when different isotopes of an element are not randomly present (e.g., when the phenyl rings of polystyrene are selectively deuterated), eq. (4.13) cannot be used to determine the coherent scattering of the sample since the scattering lengths are no longer independent of atomic position.

4.1.2 Inelastic and quasielastic dynamic structure factors

Above it was assumed that the interaction of the neutrons with the nuclei is completely elastic – there is no energy/momentum gain or loss by the neutrons due to collisions with the nuclei, allowing neglect of time/frequency dependence of the scattering. In reality neutrons will exchange energy with the materials they interact with, leading to an energy (frequency) spectrum for the scattered neutrons. This spectrum (for a given q) has two components: an elastic (zero energy exchange) component consisting of a very sharp line centered at $\hbar\omega = 0$, where ω is the *change in frequency* of the neutron due to scattering, and an inelastic component corresponding to an exchange of energy between the neutron and the material. Collisions of the neutron with the material can result in a change of vibrational quantum states of the material. In polymers, inelastic scattering due to vibrational transitions leads to broad peaks in the spectrum centered off zero. Such processes, associated with changes in the vibrational quantum state of the side chain or main chain polymer, are not relaxational and will not be discussed in this book. Rotational and translation (relaxational) motion leads to a broadening of the elastic line ($\hbar\omega = 0$). The broadening of the elastic line is referred to as quasielastic neutron scattering (QENS). For inelastic scattering the double differential cross-section is [1,2]

$$\frac{\partial^2 \sigma}{\partial \Omega \partial \omega} = \frac{k_f}{k_i} \frac{1}{2\pi} \int_{-\infty}^{\infty} \exp(-i\omega t) \left\langle \sum_{j,k}^{N} b_j b_k \exp i\mathbf{q} \cdot (\mathbf{r}_k(t) - \mathbf{r}_j(0)) \right\rangle dt. \quad (4.14)$$

As with elastic scattering, inelastic scattering has both incoherent (self) and coherent (self + distinct) components, and can be written, when one element dominates scattering, as [1,2]

$$\frac{\partial^2 \sigma}{\partial \Omega \partial \omega} = \frac{k_f}{k_i} \frac{\Delta b^2}{2\pi} \int_{-\infty}^{\infty} \exp(-i\omega t) \left\langle \sum_{k}^{N} \exp i\mathbf{q} \cdot (\mathbf{r}_k(t) - \mathbf{r}_k(0)) \right\rangle dt$$

$$+ \frac{k_f}{k_i} \frac{\langle b \rangle^2}{2\pi} \int_{-\infty}^{\infty} \exp(-i\omega t) \left\langle \sum_{j,k}^{N} \exp i\mathbf{q} \cdot (\mathbf{r}_k(t) - \mathbf{r}_j(0)) \right\rangle dt$$

$$= \frac{\partial^2 \sigma}{\partial \Omega \partial \omega}\bigg|_{\text{inc}} + \frac{\partial^2 \sigma}{\partial \Omega \partial \omega}\bigg|_{\text{coh}}. \quad (4.15)$$

When multiplied by the flux of incoming neutrons, the double differential cross-section gives the number of neutrons scattered into a solid angle element $d\Omega$ with an energy transfer $\hbar\omega$. The incoherent term depends upon the motion of the individual nuclei. The second term depends upon the static ($t = 0$) and dynamic (motional) correlation between atoms. While coherent inelastic scattering contains more information than incoherent scattering, it is more difficult to interpret. The dynamic cross-section is commonly written expressed in terms of the dynamic scattering function, or dynamic structure factor $s(q, \omega)$, yielding [1,2]

$$\frac{\partial^2 \sigma}{\partial\Omega\partial\omega}\bigg|_{\text{inc}} = N\frac{k_f}{k_i}\Delta b^2 s_{\text{inc}}(q, \omega), \qquad (4.16)$$

$$\frac{\partial^2 \sigma}{\partial\Omega\partial\omega}\bigg|_{\text{coh}} = N\frac{k_f}{k_i}\langle b\rangle^2 s_{\text{coh}}(q, \omega), \qquad (4.17)$$

where

$$s_{\text{inc}}(q, \omega) = \frac{1}{2\pi N}\int_{-\infty}^{\infty}\exp(-i\omega t)\left\langle\sum_{k}^{N}\exp i\mathbf{q}\cdot(\mathbf{r}_k(t) - \mathbf{r}_k(0))\right\rangle dt, \quad (4.18)$$

$$s_{\text{coh}}(q, \omega) = \frac{1}{2\pi N}\int_{-\infty}^{\infty}\exp(-i\omega t)\left\langle\sum_{j,k}^{N}\exp i\mathbf{q}\cdot(\mathbf{r}_k(t) - \mathbf{r}_j(0))\right\rangle dt. \quad (4.19)$$

4.1.3 Intermediate dynamic structure factors

The connection between molecular motion and neutron scattering dynamic structure factors is more transparent in the time domain. The dynamic intermediate incoherent and coherent structure factors are given by [1,2]

$$s_{\text{inc}}(q, t) = \int_{-\infty}^{\infty}\exp(i\omega t)s_{\text{inc}}(q, \omega)\,d\omega, \qquad (4.20)$$

$$s_{\text{coh}}(q, t) = \int_{-\infty}^{\infty}\exp(i\omega t)s_{\text{coh}}(q, \omega)\,d\omega, \qquad (4.21)$$

yielding

$$s_{\text{inc}}(q, t) = \frac{1}{N}\left\langle\sum_{k}^{N}\exp i\mathbf{q}\cdot(\mathbf{r}_k(t) - \mathbf{r}_k(0))\right\rangle, \qquad (4.22)$$

$$s_{\text{coh}}(q, t) = \frac{1}{N}\left\langle\sum_{j,k}^{N}\exp i\mathbf{q}\cdot(\mathbf{r}_k(t) - \mathbf{r}_j(0))\right\rangle, \qquad (4.23)$$

Table 4.1. *Neutron scattering cross-sections*
of common elements in organic polymers

Element	$\sigma_{coh} = 4\pi \langle b^2 \rangle$ (barn)	$\sigma_{coh} = 4\pi \langle \Delta b^2 \rangle$ (barn)
C	5.551	0.001
H	1.7568	80.26
D	5.592	2.05
O	4.232	0.008
N	11.01	0.007

Data from Zorn [2].

when scattering is dominated by one element. When more than one element contributes significantly to the scattering, the self (incoherent) and self + distinct (coherent) dynamic correlations for each element or element pair, respectively, will contribute differently to the total dynamic structure factor of the sample and eqs. (4.22) and (4.23) are no longer applicable to the sample as a whole but rather to each element and element pair. The incoherent and coherent dynamic structure factors of the sample are given, respectively, as a weighted (by number density and scattering length) sum of the contributions of each element and element pair.

4.1.4 Neutron scattering cross-sections and isotopic substitution

The neutron scattering cross-sections of common elements in organic polymers are shown in Table 4.1 As a result of the large incoherent scattering cross-section of hydrogen, incoherent scattering in hydrogenous materials is dominated by hydrogen atoms, allowing application of eq. (4.22) where the sum is limited to hydrogen atoms. When the Gaussian approximation for atomic displacements is valid, which applies (strictly) in the limit of small q, diffusive motion or motion in harmonic potentials [1,2], incoherent scattering in a hydrogenous polymer is given by

$$s_{inc}(q, t) = \exp\left[-\frac{q^2 \langle \Delta r_H^2(t) \rangle}{6}\right], \tag{4.24}$$

where $\langle \Delta r_H^2(t) \rangle$ is the mean-square displacement of the hydrogen atoms. The large difference in the incoherent scattering lengths of hydrogen and deuterium further affords the possibility of measuring the incoherent dynamic structure factor, and hence investigating the motion of particular polymer segments (e.g. repeat units) by selective deuteration since non-deuterated (hydrogenous) segments will dominate the incoherent spectra.

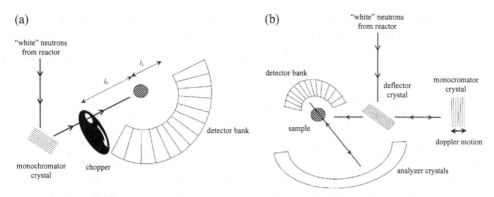

Figure 4.2 Schematic of: (a) a TOF spectrometer and (b) a backscattering spectrometer. Adapted from Zorn [2].

Table 4.1 also reveals that the coherent cross-sections of hydrogen and deuterium differ significantly, and hence selective deuteration can also be used to investigate the coherent dynamic scattering associated with particular polymer segments or even entire polymer chains. For example, use of mixtures of deuterated and protonated chains (e.g., a matrix of 10% fully protonated chains and 90% the fully deuterated chains) allows the single-chain dynamic structure factor to be determined by subtraction (with the proper transmission factors [3]) of the scattering from the matrix obtained from dynamic neutron scattering scans on the pure matrix (e.g. all chains deuterated).

4.2 Time-of-flight (TOF) and backscattering QENS

4.2.1 Methods

The neutron time-of-flight (TOF) spectrometer, illustrated schematically in Figure 4.2(a), provides information on the neutron energy change $\hbar\omega$ in a sample by measuring both the time that a neutron needs to reach the sample from a known starting point and the time after the scattering process it needs to reach the detector. In TOF measurements the instrument is surrounded by an array of detectors allowing simultaneous observation of a range of q vectors and energy transfers $\hbar\omega$. A neutron chopper in the incident beam defines the start time. From the time it takes the neutron to arrive at a detector (time of flight) its velocity and hence its final energy are determined. The initial energy E_i is typically established by a monochromator crystal or a series of choppers, allowing a single wavelength to reach the sample. The energy transfer is given by [2]

$$\hbar\omega = \left(\frac{l_1^2}{\left(l_0 - \sqrt{2E_i m_n} t_{\text{flight}}\right)^2} - 1\right) E_i, \tag{4.25}$$

where l_1 and l_0 are the distance from the chopper to the sample and the distance from the sample to the detector, respectively. The neutron momentum transfer at a given detector (scattering angle) depends on the energy transfer and is given by [2]

$$q = \frac{\sqrt{2m_n}}{\hbar}(2E_i + \hbar\omega - 2\sqrt{E_i(E_i + \hbar\omega}\cos 2\theta)^{1/2}. \tag{4.26}$$

The wavelength of the incident neutrons determines the accessible q range and energy resolution. Hotter (smaller wavelength) neutrons provide access to a larger q range, while colder neutrons yield better energy resolution, and hence access to longer relaxation times. The energy resolution of TOF spectrometers is limited to about 50 μeV, corresponding to a maximum time scale of about 40 ps.

In order to increase energy resolution and allow access to longer relaxation times backscattering spectrometers can be used. The energy resolution of a TOF spectrometer is limited by the monochromator crystal [2]. If a perfect crystal is used the spread of the selected wavelengths $\Delta\lambda/\lambda$ is determined by differentiating the Bragg condition $\lambda = 2\sin\theta/d$ yielding

$$\Delta\lambda/\lambda = \cot\theta\,\Delta\theta, \tag{4.27}$$

which becomes zero for $2\theta = 180°$. Hence, wavelength spread becomes minimal if the neutron beam is reflected by $180°$, i.e., under backscattering conditions. As shown in Figure 4.2(b), backscattered neutrons from the monochromator are scattered by the sample and scattered again by an array of analyzer crystals under backscattering conditions, finally reaching the detector. The Doppler motion of the monochromator crystal (or alternatively use of a heated crystal) allows the backscattering spectrometer to be used to study inelastic scattering. Backscattering instruments have an energy resolution on the order of 1 μeV, corresponding to about 2 ns.

4.2.2 Example application of incoherent TOF QENS to a polymer melt

The incoherent intermediate dynamic structure factor $S_{inc}(q, t)$ obtained for an unentangled poly(ethylene) melt ($C_{100}H_{202}$) at 509 K from both neutron scattering measurements and MD simulations is shown in Figure 4.3 [4]. QENS experiments were performed on a hydrogenous poly(ethylene) sample using the TOF spectrometer IN6 at the Institut Laue-Langevin in Grenoble for six momentum transfers ranging from $q = 0.8$ Å^{-1} to $q = 1.8$ Å^{-1}. MD simulations were performed using both an explicit atom (EA) model for poly(ethylene) and a united atom (UA) model where the hydrogen atoms are subsumed into the carbons. While the MD simulations yield $s_{inc}(q, t)$ directly, experimental determination of $s_{inc}(q, t)$

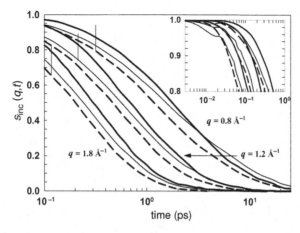

Figure 4.3   The intermediate dynamic structure factor for n-$C_{100}H_{202}$ melts as a function of time. Thick solid lines are from experimental TOF measurements, thin solid lines are from explicit atom MD simulations and dashed lines are from united atom MD simulations. The vertical lines show the lower time limit of the experiments. The inset emphasizes short time behavior. Data from Smith *et al.* [4].

requires application of eq. (4.20) and correction for the resolution factor of the instrument, i.e.

$$s_{inc}(q, t) = \frac{\int_{-\infty}^{\infty} \exp(i\omega t)s_{inc}(q, \omega)\, d\omega}{\int_{-\infty}^{\infty} \exp(i\omega t)R(\omega)\, d\omega}, \tag{4.28}$$

where $R(\omega)$ is the instrument resolution function. In order to carry out this Fourier transformation it was assumed that $s_{inc}(q, \omega)$ followed a power-law decay for high frequencies (i.e., energy transfers) beyond the resolution of the instrument. The influence of the high frequency, power-law extrapolation on the experimental data can be seen in Figure 4.3. The extrapolated experimental data miss contributions to hydrogen motion and hence decay of $s_{inc}(q, t)$ at short times from high frequency, valence angle bending modes that are captured by the MD simulations. Neverthe-less, most of the decay of $s_{inc}(q, t)$ occurs within the experimental time window and at times longer than those associated with vibrational motions, and hence the hydrogen motions responsible for the decay of $s_{inc}(q, t)$ must be due to dihedral librations and conformational transitions. Access to $s_{inc}(q, t)$ on the time scale of conformational motions over a range of momentum transfers (e.g. length scales) from these TOF QENS measurements and the excellent agreement between exper-iment and simulation for times within the experimental time window have resulted

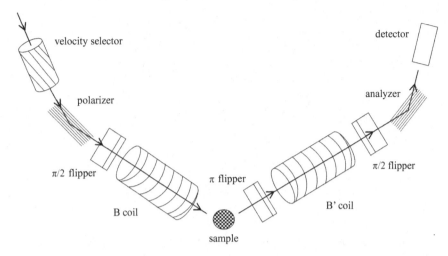

Figure 4.4 Schematic setup of an NSE spectrometer. Adapted from Zorn [2].

in important insight into the nature of conformational motions in polymer melts as described in Sec. 6.5.

4.3 Neutron spin echo (NSE) spectroscopy

4.3.1 Method

Neutron spin echo (NSE) spectroscopy has the advantages of covering the time scale of relaxation processes for polymers and furthermore yields data directly in the time domain. The experimental apparatus is shown in Figure 4.4. The principal components are the velocity selector, which preselects a wavelength spread $\Delta\lambda/\lambda$ of approximately 10%, a supermirror polarizer, which polarizes the neutron spins in the flow direction, two identical precession fields (solenoids of length m and field **H**, one before the sample, and one following the sample), two $\pi/2$ coils, and a π coil.

The first $\pi/2$ coil rotates the spin of the polarized neutrons, which precesses about the field of the first precession solenoid. The precession frequency is given by

$$\omega_L = \frac{2\mu_n H}{h}, \tag{4.29}$$

where μ_n is the magnetic moment of the neutron. The number of precessions N_0 made by a neutron is determined by ω_L (and hence the field strength) and the length of time the neutron spends in the field. The latter is determined by the neutron velocity and the length of the solenoid. After passing through the precession field neutrons lose polarization (precess to various degrees) due to their different velocities. In the absence of a sample, full polarization is regained through use of the π coil and an identical precession field. The π coil rotates the spins about a field perpendicular to the guide field. As a result, those spins that were furthest behind (largest phase lag) are now the furthest ahead, and vice versa. Consequently, after passing through the second precession field the neutrons regain 100% polarization at the second $\pi/2$ coil, which rotates the spins back into the flight direction to be observed at the analyzer [2].

When a sample is placed in the path of a neutron, the velocity of the neutron changes depending upon its exchange of energy with the sample. A given neutron will take a longer/shorter time to pass through the second precessional field than the first, thereby undergoing more/fewer precessions. As a result, complete polarization is not regained at the analyzer. The net polarization P_z for the beam scattered at a given angle ($\mathbf{q}$) is given by [2,5]

$$P_z = \int_0^\infty F(\lambda)\,d\lambda \int_{-\infty}^{+\infty} P(\lambda, d\lambda) \cos\left(\frac{2\pi N_0 \delta\lambda}{\lambda_0}\right) d(\delta\lambda). \qquad (4.30)$$

Here $F(\lambda)d\lambda$ is the wavelength spread (of the incident beam) and $P(\lambda, \delta\lambda)$ is the probability that a neutron of wavelength λ will be scattered with a wavelength change $\delta\lambda$. The wavelength change corresponds to an energy (frequency) change, yielding

$$P_z(q) = \int_0^\infty F(\lambda)\,d\lambda \frac{\int_{-\infty}^{+\infty} S_{\mathrm{coh}}(q, \omega) \cos\left(\frac{N_0 m \lambda^3}{h\lambda_0}\omega\right) d\omega}{\int_{-\infty}^{+\infty} S_{\mathrm{coh}}(q, \omega)\,d\omega}$$

$$= \int_0^\infty F(\lambda)\,d\lambda \int_{-\infty}^{+\infty} \frac{S_{\mathrm{coh}}(q, \omega)}{S_{\mathrm{coh}}(q)} \cos\left(\frac{N_0 m \lambda^3}{h\lambda_0}\omega\right) d\omega. \qquad (4.31)$$

The term $N_0 m \lambda^3 / h\lambda_0$ represents the time a neutron of wavelength λ spends in the precessional field, i.e., $t(\lambda)$. Equation (4.31) is therefore the Fourier transform of the coherent dynamic structure factor. Hence,

$$\frac{S_{\mathrm{coh}}(q, t)}{S_{\mathrm{coh}}(q)} = P_z(q). \qquad (4.32)$$

The intermediate (time domain) coherent dynamic structure factor is hence observed directly in the NSE experiment. The time scale is governed through N_0 and hence the strength of the magnetic guide field. This scale covers 10^{-11} seconds to about 10^{-7} seconds [2,5]. It is also worth noting that improved instrumentation including

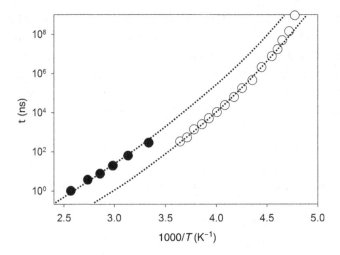

Figure 4.5 Relaxation times for the α-relaxation process in a poly(isobutylene) melt. Filled circles are from NSE measurements, open circles from dielectric measurements. The dotted lines show the temperature dependence of the viscosity. Data from Richter *et al.* [7].

multidetectors allows NSE to be used to probe the q- and t-dependence of the incoherent intermediate scattering function [6].

4.3.2 Example application of NSE spectroscopy to a polymer melt

Relaxation times for $s_{\mathrm{coh}}(q, t)/s_{\mathrm{coh}}(q)$ determined from NSE experiments performed on a perdeuterated poly(isobutylene) melt are shown in Figure 4.5. The NSE experiments were performed using the IN11 spectrometer at the Institut Laue-Langevin in Grenoble. The data shown are for $q = 1.0\ \text{Å}^{-1}$, corresponding to the first peak in the static structure factor of poly(isobutylene). The relaxation times were determined from KWW fits of the experimental $s_{\mathrm{coh}}(q, t)/s_{\mathrm{coh}}(q)$ data. Also shown are the relaxation times for the α-relaxation process in the same sample determined from dielectric relaxation measurements. Reasonable correspondence of the relaxation times for the two experimental probes can be seen. Furthermore, the lines in Figure 4.5 show the temperature dependence of the viscosity of poly(isobutylene). The correspondence between the temperature dependence of the NSE relaxation times, dielectric relaxation times and that of the viscosity supports the contention that NSE measurements at this q value (first peak in the static structure factor) are probing atomic motions that are fundamentally responsible for the relaxation of the polymer on length scales ranging from segmental to the terminal relaxation of the polymer chain.

References

[1] J. S. Higgins and H. C. Benoît, *Polymers and Neutron Scattering* (New York: Oxford University Press, 1994).

[2] R. Zorn in *Neutrons, X-Rays and Light*, edited by P. Linder and T. Zemb (New York: Elsevier, 2002), Chapter 10.

[3] B. K. Annis, O. Borodin, G. D. Smith, *et al.*, *J. Chem. Phys.*, **115**, 10998 (2001).

[4] G. D. Smith, W. Paul, D. Y. Yoon, *et al.*, *J. Chem. Phys.*, **107**, 4751 (1997).

[5] B. Ewen and D. Richter, *Adv. in Polym. Sci.*, **34**, 1 (1997).

[6] J. Colmenero, A. Arbe, D. Richter, B. Farago, and M. Monkenbusch in *Neutron Spin Echo Spectroscopy: Basics, Trends and Applications* (Lecture Notes in Physics), edited by F. Mezei, C. Pappas and T. Gutberlet (Berlin: Springer-Verlag, 2003) pp. 268–279.

[7] D. Richter, A. Arbe, J. Colmenero, *et al.*, *Macromolecules* **31**, 1133 (1998).

5

Molecular dynamics (MD) simulations
of amorphous polymers

Atomistic MD simulations are "computer experiments" in which realistic trajectories of the time dependent positions of atoms that make up the simulated material
are generated. From these simulations time correlation functions directly associated
with relaxation processes of interest can be determined. Hence, MD simulations
are literally applied statistical mechanics. MD simulations of amorphous polymers provide a molecular level picture of their structure and dynamics and thereby
insight into relaxation mechanisms not available from experiments. The simulator
can also explore relaxation mechanisms by performing parametric investigations
of the role of polymer structure and energetics (e.g., rotational energy barriers) on
relaxations that cannot be mimicked experimentally. Perhaps most importantly, synergistic relationships between simulators and experimentalists can, as demonstrated
throughout this book, result in new insights into polymer dynamics and relaxation
processes not accessible to either approach alone. This chapter is intended to introduce the non-expert to the concepts, applications, and limitations of atomistic MD
simulations of bulk amorphous polymers, particularly with regard to the investigation of polymer dynamics and relaxation processes. A more detailed description
of MD simulation methods can be found in several excellent textbooks [1,2] and
compilations [3–6].

5.1 A brief history of atomistic MD simulations of amorphous polymers

MD simulations were first applied to biological macromolecules (proteins) in the
1970s [7,8], motivated by efforts to understand single-molecule folding. Shortly
thereafter the method was applied to the study of the dynamics of a single synthetic
polymer chain (poly(ethylene)) [9]. Application of MD to bulk amorphous polymers was long considered not to be viable due to both the computational cost and
the clear complexity of motion within a bulk polymer melt which seemed intuitively
too difficult to reproduce by simulations. Furthermore, it was not clearly realized

that extremely valuable insight into polymer dynamics and relaxation behavior can be obtained from simulations of even modest length (e.g., a nanosecond or less) due to the fact that much of the experimentally observed behavior in bulk polymers occurs on very long (e.g., a second or longer) time scales. Nevertheless, after an early effort by Weber and Helfand [9], application of MD simulation methods joined the mainstream of computational tools used in studying bulk polymers with the seminal work of Rigby and Roe [10,11]. By simplifying the structure of alkane polymers and reducing force constants associated with the highest frequency motions, thereby allowing an increase in the integration time step, they were able to generate configurations of bulk polymers and generate simulation trajectories of sufficient length to allow the determination of both static and dynamic properties. MD simulations have become an indispensable tool for polymer scientists as witnessed by the large number of papers and books utilizing the method and the increasing use of MD simulations by non-experts.

5.2 The mechanics of MD simulations

The classical atomistic MD simulation method invokes a simple yet fundamental set of rules for generating a series of representative configurations (atomic positions and velocities) of a collection of atoms that trace the time evolution of that system in the statistical mechanical ensemble of interest. In each valid classical statistical mechanical ensemble, there is a quantity that is a "constant of motion", i.e., is equal for all statistically mechanically valid configurations of the material at the established thermodynamic conditions. The NVE (constant Number of particles, Volume, and Energy) ensemble, while not often utilized in practice, forms the basis of understanding of how MD simulations are performed. In the NVE ensemble, this constant, referred to as the Hamiltonian H, is simply the total energy of the ensemble of atoms, given as

$$H(\mathbf{r}, \mathbf{p}) = K(\mathbf{p}) + V(\mathbf{r}) = E = \text{constant} \tag{5.1}$$

where $\mathbf{r}$ represents the Cartesian coordinates of all N atoms in the ensemble, $\mathbf{p}$ is their momentum, $K(\mathbf{p}) = \frac{1}{2}\sum_{i=1}^{N} p_i^2/m_i$ is the total kinetic energy, and $V(\mathbf{r})$ is the potential energy. The potential energy depends upon the position of all atoms (force centers) within the ensemble and is discussed in detail below. Equation (5.1) yields Hamilton's equations of motion,

$$\frac{d\mathbf{r}_i(t)}{dt} = \frac{\partial H(\mathbf{r}, \mathbf{p})}{\partial \mathbf{p}_i} = \frac{\mathbf{p}_i(t)}{m_i}$$

$$\frac{d\mathbf{p}_i(t)}{dt} = -\frac{\partial H(\mathbf{r}, \mathbf{p})}{\partial \mathbf{r}_i} = -\frac{\partial V(\mathbf{r})}{\partial r_i} = \mathbf{F}_i(\mathbf{r}, t) \qquad 1 \leq i \leq N, \tag{5.2}$$

where $\mathbf{F}_i(\mathbf{r}, t)$ is the force acting on atom i at time t due to the presence of the $N - 1$ other atoms in the ensemble. This force depends upon the position of all particles, and hence implicitly on time. The classical equations of motion are integrated numerically, yielding an update of the velocities and positions of the atom and hence the evolution of the positions (and velocities) of the atoms in a microcanonical (NVE) ensemble. The primary advantage of the MD simulation method over other methods (e.g., Monte Carlo methods) for sampling statistical mechanical phase space is that the dynamical evolution of the configuration, i.e., the trajectory, given as the atomic positions as a function of time, corresponds to the real dynamics of the system. Stated otherwise, the dynamics of the system obtained from MD simulation mimics, as accurately as the classical force field represents real forces, the dynamics of a real material.

Experimental studies of polymers are not conducted under NVE conditions, but rather under NVT (constant number of atoms, volume and temperature) or more commonly NPT (constant number of atoms, pressure and temperature) conditions. NVT conditions correspond to the statistical mechanical canonical ensemble and NVP conditions to the isothermal–isobaric ensemble. In practice, simulations are commonly carried out for the sake of convenience under NVT conditions once the volume corresponding to the pressure of interest is established. MD simulations can mimic NVT and NPT conditions through use of extended ensemble techniques [1]. These straightforward methods introduce additional degrees of freedom associated with the temperature or temperature and volume of the simulation cell. These degrees of freedom have an associated mass and velocity and their "position" and velocity are updated in the numerical integration along with those of the atoms themselves. The force on these degrees of freedom is a function of the deviation of the system from the desired temperature or temperature and pressure.

5.2.1 Limitations of atomistic MD simulations

MD simulations of polymers are limited by three primary factors. The first is the quality of the classical force field, i.e., the ability of the force field to reproduce the forces in real polymeric materials. Force fields are discussed in detail below. The second is the time step used in the numerical integration of the equations of motion which is very small, typically on the order of a femtosecond, due to the high frequency of atomic motions associated with covalent bonds, valence bends, and torsional degrees of freedom. The atoms of the ensemble (system) move far enough on a femtosecond time scale that their forces change sufficiently to require updating, which necessitates calculation of interatomic distances. The third limiting factor is the scaling of computational costs with the size of the ensemble (number of atoms N). Since the calculation of forces on each atom at each time step, which involves contributions from bonded (bond, bend, torsion) and non-bonded (van der Waals and electrostatic) interactions, is the resource intensive step in MD simulations, simulation costs scale at best linearly, and often more strongly, with N since long-range electrostatic interactions must often be taken into account (see below). The consequence of the small integration time step and linear (or stronger) scaling in computational cost with system size is that MD simulations of bulk polymers are typically limited to systems on the order of a few nanometers and times on the order of nanoseconds, although longer simulations have been performed.

Fortunately, as demonstrated throughout this book, the time limitation of MD simulations (at longest microseconds) does not preclude extraction of dynamical information directly relevant to understanding relaxation processes in polymers. This is due to the fact that relaxation processes move into the time/frequency window accessible to MD simulations as temperature increases, as witnessed by the many direct experimental measurements (e.g., dielectric relaxation, NMR spin-lattice relaxation, neutron scattering) of short time scale/high frequency polymer dynamics. Furthermore, the limited length scales accessible to MD simulations are not typically a problem for the study of segmental relaxation in amorphous polymers, where all important structural and dynamical correlations appear to occur on relatively small length scales (a few nanometers). The use of periodic boundary conditions, as shown in Figure 5.1, eliminates the effects of boundaries/surfaces that would strongly perturb the dynamics of the small systems typically utilized in simulation studies of amorphous polymers. The length scale limitations of MD simulations are important, however, when long length scale dynamics, such as terminal relaxation in entangled polymers, are of interest, or for polymers where heterogeneous structure, such as in semi-crystalline polymers, on length scales larger than can be accessed by MD simulations, is present. Finally, near or below the glass transition temperature relaxation times become too long to be observed directly via MD simulations. However, even here MD simulations can provide important insight into mechanisms of relaxation processes as discussed extensively in Sec. 8.2.

5.2.2 Advanced MD simulation methods

Advanced MD simulation methods have the goal of accelerating the sampling of phase space within the ensemble of interest. The advanced methods that generate real dynamical trajectories are of the greatest interest for the study of dynamics and relaxation processes. The most important acceleration methods applied in atomistic MD simulations of polymers carried out to investigate relaxation behavior are constraint methods, multiple time step integration, and mesh-based Ewald methods. Constraint methods [12] allow the fixing, or constraining, of high frequency motions, particularly those associated with covalent bonds, whose motion does not

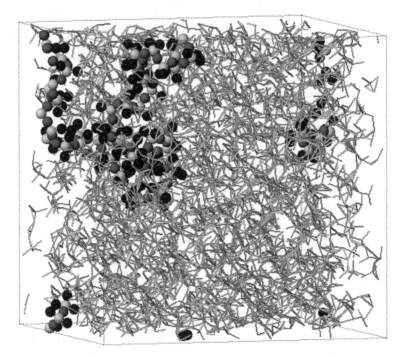

Figure 5.1 A snapshot from an atomistic MD simulation trajectory of a poly(dimethyl siloxane) melt. A space filling representation of a single chain is shown in order to illustrate periodic boundary conditions. The simulation cell is approximately 5 nm × 5 nm × 5 nm and contains approximately 12 500 atoms.

significantly influence the segmental and large scale dynamics of the system. Constraint of bond lengths allows use of a larger time integration step (femtosecond as opposed to subfemtosecond). Unfortunately, other high frequency motions such as valence bends and obviously torsional motions cannot be constrained without significantly influencing the segmental dynamics of the polymer. Multiple time step integration [13] allows relatively slowly varying forces, particularly those due to non-bonded and electrostatic interactions, that are computationally expensive, to be updated less frequently than rapidly varying forces that are relatively computationally cheap (those associated with bonds, bends, and torsions). Meshed-based Ewald methods [14] allow efficient computation of long range electrostatic interactions that are important in all polar polymers, providing $\sim N \log N$ scaling for these interactions, an important improvement over conventional Ewald ($\sim N^{3/2}$) methods for large ($N > 1000$) systems. Finally, it is worth mentioning that modern parallel computing methods can be applied to MD simulations of polymers. While parallel computing can extend the time scales accessible to MD simulations of amorphous polymers, the systems of interest are often sufficiently small

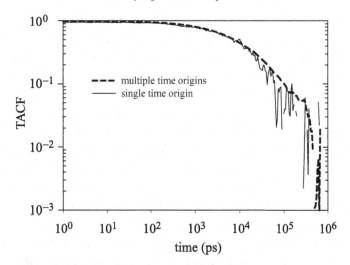

Figure 5.2 Torsional autocorrelation function (TACF, eq. (5.3)) from a 1 μs simulation of a 1,4-poly(butadiene) melt of 40 chains comprising 30 repeat units at 213 K. The influence of multiple time origin averaging on the TACF can be seen clearly. Even with multiple time origin averaging, statistical uncertainties become comparable to the value of the autocorrelation itself for times longer than about 20% of the trajectory length.

(<10 000 atoms) that the communication overhead associated with parallel computation soon becomes dominant, limiting parallel computation to relatively few processors (e.g., <20). Massively parallel computations (hundreds of processes) are better suited to extending the length scale accessible to MD simulations, and this is not typically the limiting factor in simulations of amorphous polymers as discussed above.

5.3 Studying relaxation processes using atomistic MD simulations

MD simulations generate two types of information: averages over the trajectory, corresponding to static (ensemble average) properties, and dynamical properties. The latter are of primary interest in the study of relaxation processes. As shown in the other chapters of Part I of this book, each experimental measure of relaxation in polymers has a corresponding underlying time correlation function (e.g., dipole moment, C–H vector, density fluctuations). Each of these time correlation functions can be straightforwardly extracted from an MD simulation trajectory. In order to improve the statistics of the time correlation functions, multiple time origins are typically employed, as illustrated in Figure 5.2. As an MD simulation proceeds (in time), the system becomes increasingly decorrelated from its original configuration, i.e., structural, orientational, and conformational relaxation has occurred. It is possible to utilize points (in time) along the trajectory as new $t = 0$ points in calculating the time correlation functions. In practice, new time origins are typically used much more frequently than pertinent relaxation time(s) of the system. This approach provides no additional improvement of the statistics of the time autocorrelation functions but is convenient since it does not require *a priori* knowledge of the relaxation times. While the multiple time origin approach can improve statistics dramatically for times shorter than the length of the trajectory, as illustrated in Figure 5.2, this approach provides no improvement as t approaches the length of the trajectory. Consequently, dynamic data are often shown only for times much shorter (an order of magnitude or more depending on the size of the system) than the actual length of the trajectory.

While MD simulations can directly monitor the fundamental correlation functions associated with measured relaxations, they have the additional advantage of allowing detailed examination of the conformational motions that foment relaxation processes in polymers. An example is shown in Figure 5.2 for a 1,4-poly(butadiene) melt. Here, the torsional autocorrelation function,

$$\text{TACF}(t) = \frac{\langle |\theta(t)|\, |\theta(0)| \rangle - \langle |\theta(0)| \rangle^2}{\langle |\theta(0)|^2 \rangle - \langle |\theta(0)| \rangle^2} \tag{5.3}$$

is shown for backbone C_{sp^2}–C_{sp^3}–C_{sp^3}–C_{sp^2} (alkyl) dihedrals of 1,4-poly(butadiene) chains, where $|\theta(t)|$ is the (absolute) value of a conformational angle for a given dihedral at time t and the ensemble average is taken over all alkyl dihedrals. The decay of the TACF, first used by Takeuchi and Okazaki [15], which occurs as backbone dihedrals explore conformational space through conformational transitions, is closely related to other measures of segmental relaxation as probed by neutron scattering, NMR T_1 and dielectric relaxation as shown in Chapters 6 and 8, but is a more direct measure of conformational relaxation.

5.4 Classical atomistic force fields

Reproduction of the behavior and properties of polymers in MD simulations requires accurate potentials and even qualitative conclusions drawn from simulations employing inaccurate or non-validated potentials are problematical. When interest lies in reproducing the dynamic and relaxation properties of a polymer, the potential must accurately represent the molecular geometry, non-bonded interactions, and conformational energetics of the polymer of interest. The simple representation of the classical potential energy discussed below has been found to work remarkably well for these properties for many polymers.

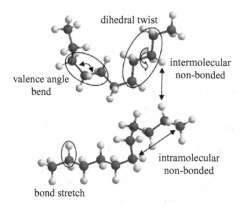

Figure 5.3 Bonded and non-bonded (two-body) interactions arising in a polymer system as represented by a simple classical potential.

5.4.1 Form of the potential

The classical force field represents the total potential energy of an ensemble of atoms $V(\mathbf{r})$ with positions given by the vector $\mathbf{r}$ as a sum of non-bonded interactions $V^{\mathrm{NB}}(\mathbf{r})$ and energy contributions due to all bond, valence bend, and dihedral interactions:

$$V(\mathbf{r}) = V^{\mathrm{NB}}(\mathbf{r}) + \sum_{\mathrm{bonds}} V^{\mathrm{BOND}}(r_{ij}) + \sum_{\mathrm{bends}} V^{\mathrm{BEND}}(\theta_{ijk}) + \sum_{\mathrm{dihedrals}} V^{\mathrm{TORS}}(\varphi_{ijkl}).$$

(5.4)

The various interactions are illustrated in Figure 5.3. The dihedral term also includes four-center improper torsion or out-of-plane bending interactions that occur at sp^2 hybridized centers.

Commonly, the non-bonded energy, $V^{\mathrm{NB}}(\mathbf{r})$, consists of a sum of two-body repulsion and dispersion energy terms between atoms i and j represented by the Buckingham (exponential-6) potential, the energy due to the interactions between fixed partial atomic or ionic charges (Coulomb interaction), and the energy due to many-body polarization effects:

$$V^{\mathrm{NB}}(\mathbf{r}) = V^{\mathrm{POL}}(\mathbf{r}) + \frac{1}{2} \sum_{i,j=1}^{N} A_{ij} \exp(-B_{ij} r_{ij}) - \frac{C_{ij}}{r_{ij}^6} + \frac{q_i q_j}{4\pi \varepsilon_0 r_{ij}}.$$

(5.5)

In addition to the exponential-6 form, the Lennard–Jones form of the dispersion–repulsion interaction,

$$V^{\mathrm{DIS-REP}}(r_{ij}) = \frac{A_{ij}}{r_{ij}^{12}} - \frac{C_{ij}}{r_{ij}^6} = 4\varepsilon \left[\left(\frac{\sigma}{r_{ij}}\right)^{12} - \left(\frac{\sigma}{r_{ij}}\right)^6 \right]$$

(5.6)

is commonly used, although this form tends to yield a poorer (too stiff) description of repulsion. Non-bonded interactions are typically included between all atoms of different molecules and between atoms of the same molecule separated by more than two bonds (see Figure 5.3). Repulsion parameters have the shortest range and typically become negligible at 1.5σ, where σ (see eq. (5.6)) is around 4 Å. Dispersion parameters are longer range than the repulsion parameters requiring cutoff distances of 2.5σ or greater. The Coulomb term is long range, necessitating use of special summing methods as discussed above.

A further complication arises in cases where many-body dipole polarization needs to be taken into account explicitly. This can occur for highly polar polymers, charged polymers, or polymers in the presence of charged species (e.g., salts or surfaces). The potential energy due to dipole polarization is not pair-wise additive and is given by a sum of the interaction energy between the induced dipoles μ_i and the electric field $\mathbf{E}_i^0$ at atom i generated by the permanent charges in the system (q_i), the interaction energy between the induced dipoles, and the energy required to induce the dipole moments [16]. Accounting for many-body polarization typically increases the computational cost by a factor of at least 3. In contrast, while most polymers are sufficiently polar that Coulomb interactions must be accurately represented in order to adequately reproduce intra- and intermolecular interactions, and in some cases many-body polarization effects must be accounted for, there are important exceptions, particularly polyolefins and perfluoroalkanes. For such (essentially) non-polar polymers it is often possible to "subsume" atoms (i.e., hydrogen or fluorine atoms) into their attached carbons, resulting in "united atom" potentials which are about an order of magnitude more computationally efficient than the corresponding (non-polar) fully atomistic representation.

In contrasting the form of the potential for polymers with those for other materials, the nature of bonding in polymers becomes apparent. In polymers, relatively strong covalent bonds and valence bends serve primarily to define the geometry of the molecule. The thermodynamic and dynamic properties of polymers are primarily determined by much weaker/softer degrees of freedom, namely those associated with torsions and non-bonded interactions. Hence relatively weak (and consequently difficult to parameterize) torsional and repulsion/dispersion parameters must be determined with great accuracy in potentials for polymers. However, this separation of scales of interaction strengths (strong intramolecular covalent bonding, weak intermolecular bonding) has the advantage of allowing many-body interactions, which often must be treated through explicit many-body non-bonded terms in simulations of other classes of materials, to be treated much more efficiently as separate intramolecular bonded interactions in polymers.

5.4.2 Sources of atomistic potentials

In general, force fields for polymers can be divided into three categories: force fields parameterized based upon a broad training set of molecules such as small organic molecules, peptides, or amino acids; these include AMBER [17], COMPASS [18], OPLS-AA [19] and CHARMM [20]; generic potentials such as DREIDING [21] and UNIVERSAL [22] that are not parameterized to reproduce properties of any particular set of molecules; and specialized force fields carefully parameterized to reproduce properties of a specific polymer. Parameterized force fields can work well within the class of molecules upon which they have been parameterized. However, when the force field parameters are utilized for compounds similar to those in the original training set but not contained in that training set significant errors can appear and the quality of force field predictions is often no better than that of the generic force fields [23]. Similar behavior can be expected when parameterized force fields are transferred to new classes of compounds. In choosing a potential for polymer simulations both the quality and the transferability of the potential need to be considered.

In order to judge the quality of existing force fields for a polymer of interest, or to undertake the demanding task of parameterizing or partially parameterizing a new force field, one requires data against which the force field parameters (or a subset thereof) can be tested and, if necessary, fit. There are two primary sources for such data: experiment and ab initio quantum chemistry calculations. Experimentally measured structural, thermodynamic, and spectroscopic data for condensed phases (liquid and/or crystal) of the polymer of interest or closely related compounds are particularly useful in force field parameterization and validation. High-level quantum chemistry calculations are the best source of molecular level information for force field parameterization. While such calculations are not yet possible on high molecular weight polymers they are feasible on small molecules representative of polymer repeat units and oligomers as well as on molecular clusters that reproduce interactions between segments of polymers or the interaction of these with surfaces, solvents, ions, etc. These calculations can provide the molecular geometries, partial charges, polarizabilities, conformational energy surface, and intermolecular non-bonded interactions critical for accurate prediction of structural, thermodynamic and dynamic properties of polymers [24,25].

References

[1] M. P. Allen and D. J. Tildesley, *Computer Simulation of Liquids* (New York: Oxford University Press, 1987).
[2] D. Frenkel and B. Smit, *Understanding Molecular Simulation*, Computational Science Series, Vol. 1 (San Diego: Academic Press, 2002).
[3] K. Binder (ed.), *Monte Carlo and Molecular Dynamics Simulations in Polymer Science* (New York: Oxford University Press, 1995).
[4] M. Kotelyanski and D. N. Theodorou (eds.), *Simulation Methods for Polymers* (New York: Marcel Dekker, 2004).
[5] V. Galiatsos, *Molecular Simulation Methods for Predicting Polymer Properties* (New York: Wiley, 2005).

[6] P. Pasini, C. Zannoni, and S. Žumer (eds.), *Computer Simulations of Liquid Crystals and Polymers* (Dordrecht, The Netherlands: Kluwer Academic, 2005).

[7] J. A. McCammon, B. R. Gelin, and M. Karplus, *Nature*, **267**, 585 (1977).

[8] F. R. N. Gurd and J. M. Rothgeb, *Adv. Protein Chem.*, **33**, 73 (1979).

[9] T. A. Weber and E. Helfand, *J. Chem. Phys.*, **71**, 4760 (1979).

[10] D. Rigby and R. J. Roe, *J. Chem. Phys.*, **87**, 7285 (1987).

[11] D. Rigby and R. J. Roe, *J. Chem. Phys.*, **89**, 5280 (1988).

[12] J. P. Ryckaert, G. Ciccotti, and H. J. C. Berendsen, *J. Comput. Phys.*, **23**, 327 (1977).

[13] M. Tuckerman, G. J. Martyna, and B. J. Berne, *J. Chem. Phys.*, **97**, 1990 (1992).

[14] A. Toukmaji, C. Sagui, J. Board, and T. Darden, *J. Chem. Phys.*, **113**, 10912 (2000).

[15] H. Takeuchi and K. Okazaki, *J. Chem. Phys.*, **92**, 5643 (1990).

[16] T. M. Nymand and P. Linse, *J. Chem. Phys.*, **112**, 6152 (2000).

[17] W. D. Cornell, P. Cieplak, C. I. Bayly *et al.*, *J. Am. Chem. Soc.*, **117**, 5179 (1995).

[18] H. Sun, *J. Phys. Chem. B*, **102**, 7338 (1998).

[19] W. L. Jorgensen, D. S. Maxwell, and J. Tirado-Rives, *J. Am. Chem. Soc.*, **118**, 11225 (1996).

[20] A. D. MacKerell Jr., D. Bashford, M. Bellot *et al.*, *J. Phys. Chem. B*, **102**, 3586 (1998).

[21] S. L. Mayo, B. D. Olafson, and W. A. Goddard III, *J. Phys. Chem.*, **94**, 8897 (1990).

[22] A. K. Rappé, C. J. Casewit, K. S. Colwell, W. A. Goddard, and W. M. Skiff, *J. Am. Chem. Soc.*, **114**, 10024 (1992).

[23] F. Sato, S. Hojo, and H. Sun, *J. Phys. Chem. A*, **107**, 248 (2003).

[24] G. D. Smith and O. Borodin, in *Molecular Simulation Methods for Predicting Polymer Properties*, edited by V. Galiatsatos (New York: Wiley, 2005), Chapter 2.

[25] G. D. Smith, in *Handbook of Materials Modeling*, edited by S. Yip (New York: Springer, 2005), Chapter 9.2.

Part II

Amorphous polymers

Amorphous polymers represent in one sense the simplest class of polymers in that they have no intermediate microstructure between that of the macroscopic specimen and the molecular level. Very often intermediate scale microstructures have an influence on relaxational behavior and thus wholly amorphous polymers are a good baseline to which more complicated structures can be compared. In another sense the amorphous state is exceedingly complicated. At an atomistic level it is difficult to describe or model an interpenetrating collection of chains, each pursuing a more or less random spatial trajectory. In a given amorphous polymer usually more than one region of relaxation in time and temperature can be detected. The term amorphous implies that, if chemical stability is adequate, at some high temperature the polymer exists as a viscous liquid or melt and that at sufficiently low temperature vitrification will take place. The relaxation process associated with the glass transition region is thus very prominent. Usually, however, one or more additional relaxation processes are found at temperatures below the glass transition temperature. The glass transition relaxation region and subglass relaxations each have their own set of common characteristics or "signatures." MD simulations are now able to accomplish the modeling of the structure of amorphous polymers and over time trajectories long enough to be useful in interpreting the relaxation processes.

A word about the nomenclature used in denoting multiple relaxation processes in the same polymer is in order. It has become customary to use lower case Greek letters with the convention that the highest temperature process observed isochronally or the lowest frequency one observed isothermally is assigned α. Then β, γ and so on are invoked for descending temperatures or ascending frequencies. One given polymer might have two relaxation processes and another might have three or even more. There is no guarantee that an α process in one polymer refers to a process of similar type or origin in another polymer that has several identifiable processes. It is best at the outset of any given discussion to use descriptive modifiers such as the "α primary relaxation" or the "β subglass relaxation." Every effort has been made here to observe these precautions.

6

The primary transition region

6.1 Mechanical relaxation

6.1.1 Scope of relaxation

Although the primary transition temperature in a given polymer tends to be similar whether studied mechanically or dielectrically the characteristics in detail are quite different. The mechanical modulus eventually falls to zero with time at higher temperatures. If the molecular weight is above the entanglement length, the rubbery plateau region intervenes before the viscous flow region and the ensuing complete decay toward zero. However, the modulus in the plateau region is orders of magnitude less than the glassy modulus. Thus the mechanical relaxation is characterized by a large modulus change that is best represented by a logarithmic scale for modulus (as in Figure 1.9 for example). Because of the large range over which the modulus decay may be measured the time scale is also extremely wide. It is actually far too wide to be captured directly experimentally. In fact, in a single apparatus, mechanical measurements are usually restricted to a few decades of time or frequency. However, by measuring time or frequency responses at a number of temperatures various parts of the overall relaxation from the glassy to the rubbery state may be captured. Figure 6.1 shows a particularly elegant set of such data [1] in which the storage compliance of poly(n-octyl methacrylate) is plotted against log frequency at a number of temperatures. The complete transition region may then be described in alternative ways.

One method of unifying the data is simply to choose temperature as the independent variable and to plot the decay or retardation versus the temperature as a family of constant time or frequency curves. These are *isochronal scans*. The data of Figure 6.1 are replotted in Figure 6.2 in isochronal form. The measurements were not carried out over the entire temperature range at all frequencies. However, notice that any one of the frequencies would be sufficient to trace out the entire transition region if temperature were varied over the appropriate range. It is a general

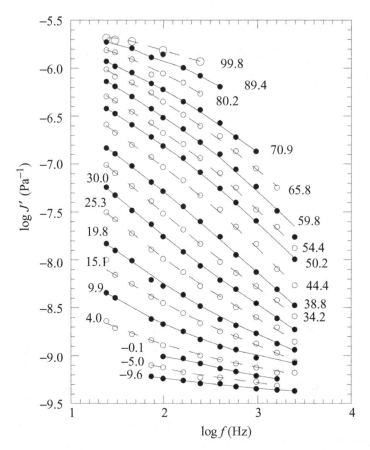

Figure 6.1 Poly(n-octyl methacrylate): dynamic storage modulus J' at the temperatures shown (°C). Data from Ferry [1].

proposition that isochronal temperature scans can sweep through an entire transition region, while a single isothermal frequency scan is often not broad enough to capture an entire relaxation process. The independently measured volumetric glass transition is indicated on Figure 6.2. It corresponds to the onset of the softening process at low frequencies.

6.1.2 Time–temperature superposition

Perusal of Figure 6.1 seems to indicate that the horizontal distance (along the log f axis) between adjacent curves, which overlap strongly in the log J' direction, is at least *approximately* constant. This suggests that a single curve might result from translations or horizontal shifts along the log f axis. This conjecture is given more quantitative substance in Figure 6.3, where all of the curves have been shifted to a

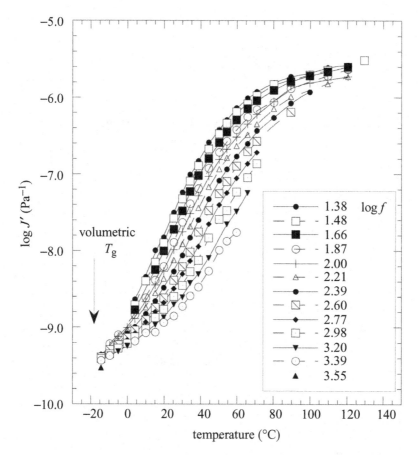

Figure 6.2 Poly(n-octyl methacrylate): dynamic storage modulus J' plotted isochronally vs. temperature at the frequencies (Hz) indicated. From the data of Figure 6.1.

common basis corresponding to 30 °C. It may be seen that a reasonable case can be made in this instance that the curves differ only by the horizontal log f shift. Such a composite curve that results from *time–temperature superposition* is often called a *master curve*.

One way to view this behavior is in terms of the distribution of relaxation or retardation times. In eq. (1.69), the distribution of retardation times for the dynamic compliance was written as

$$J^*(i\omega) = J_u + \int_{\ln \tau = -\infty}^{+\infty} \overline{L}(\tau) \left(\frac{1}{1 + i\omega\tau} \right) d\ln \tau. \tag{6.1}$$

If it is supposed that $L(\tau)$ depends on temperature *only* through a single temperature dependent parameter, $\tau_0 = \tau_0(T)$, such that $\overline{L}(\tau)$ is of the form $\overline{L}(\tau/\tau_0)$, then

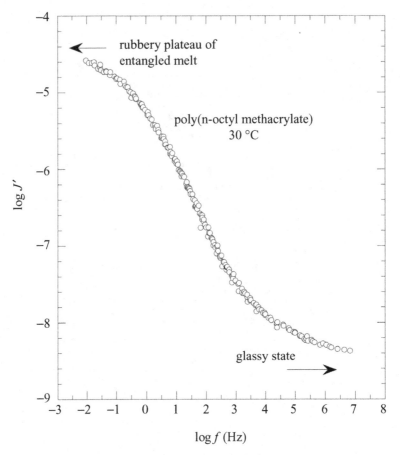

Figure 6.3 Time–temperature reduction of the data of Figure 6.1 to 30 °C.

eq. (6.1) can be written in terms of a *dimensionless reduced variable*, $\bar{\tau} = \tau/\tau_0$ as

$$J^*(i\omega) = J_u + \int_{\ln\bar{\tau}=-\infty}^{+\infty} \overline{L}(\bar{\tau}) \left(\frac{1}{1+i(\omega\tau_0)\bar{\tau}}\right) d\ln\bar{\tau}. \tag{6.2}$$

Thus, by the above conjecture, the only temperature dependence of $J^*(i\omega)$ is through τ_0. It is clear that compensatory changes in frequency and temperature can be made via keeping $\omega\tau_0$ constant. That is, at two temperatures, T_1, T_2, $J^*(\omega_1)$ will be equal to $J^*(\omega_2)$ if $\omega_1\tau_0(T_1) = \omega_2\tau_0(T_2)$. The frequency shift required to bring J^* into coincidence will be

$$\log\omega_1 - \log\omega_2 = \log f_1 - \log f_2 = \log \tau_0(T_2) - \log \tau_0(T_1). \tag{6.3}$$

A distribution of retardation or relaxation times with the above property of being capable of being written in terms of reduced variable $\overline{\tau} = \tau/\tau_0$ could be said to have a *magnitude* and *shape* that is independent of temperature and the only effect of temperature is to shift the function along $\log \tau$ and therefore the effect on the dynamic property (e.g., J^*) is to shift it along $\log f$.

It is instructive in the above context to examine the empirical representations of the dielectric constant that were introduced in Chapter 2. The Cole–Cole (eq. (2.38)), Davidson–Cole (eq. (2.41)) and HN (eq. (2.42)) equations are all formulated in terms of $\omega\tau_0$ (and via eq. (2.36) so is the distribution function, $\overline{F}(\tau)$). The requirement for applicability of time–temperature superposition explicitly becomes the requirement that the shape parameters, α and/or β, and the magnitude parameters, ε_r and ε_u, be independent of temperature. In practice, it is common to relax the requirement that the magnitude parameters be temperature independent by adopting assumptions about their behavior. No corrections were made in constructing Figure 6.3.

The practical utility of the master curve lies in the following. The actual data for each experiment can be recovered from the master curve by inverse shifting. But most importantly, the behavior of the dynamic function over the *entire* frequency range of the master plot can be inferred for *each* temperature. Thus a reasonable means for making behavior predictions far outside the experimentally accessible time or frequency region presents itself.

Finally, a rough generalization can be made that time–temperature superposition is often successful in generating an apparently acceptable master curve for mechanical relaxation in the primary transition region in amorphous polymers. However, this must be accompanied by the caveat that in the case of broadband dielectric measurements, for example, where shapes in the glass transition region can be characterized, the shapes do show temperature dependence. There are many other instances and situations where the assumptions about the temperature independence of the shape of the relaxation time distribution can be seen to be only a rough approximation. In fact, in mechanical measurements it has been observed in the transition region that in approaching the rubbery plateau region, where viscous effects begin to come into play, the temperature dependence can be quite different from the higher frequency, lower temperature portion of the transition region [2,3,4].

6.1.3 Temperature behavior of shift factors and relaxation times

The master curve embodied in Figure 6.3 was constructed from a finite set of experimental curves. Thus a set of shift factors exist for the temperatures of the experiments. If the shift factors were parameterized as a continuous temperature function

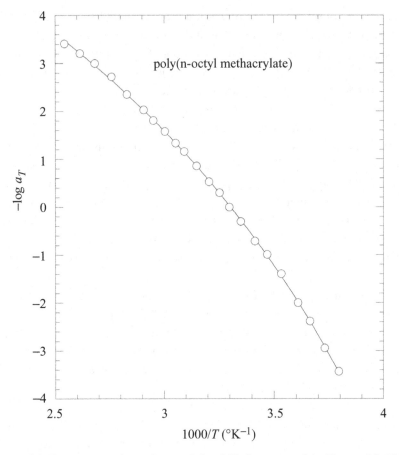

Figure 6.4 Temperature dependence of the shift factors used in Figure 6.3. The solid curve is a fit of the VF equation.

then dynamic data over the entire time/frequency region at *any* temperature could be inferred. According to the discussion of the previous section, time–temperature superposition is intimately associated with the temperature behavior of a central relaxation time, τ_0. Relaxation times in general are associated with rate processes and, as such, suggest a temperature dependence appropriate to an activated process. A frequency shift factor from temperature T to the reference temperature T_0 is defined as $\log a_T = \log f(T) - \log f(T_0)$. In Figure 6.4, $-\log a_T$ is plotted versus $1/T$ for the data of Figure 6.3. It may be seen that the figure to some degree resembles a classical straight-line Arrhenius plot whose slope is proportional to an activation energy. However, it is apparent that the plot is not a straight line and distinct curvature is seen. The slope magnitude increases as $1/T$ increases or therefore as T decreases. This non-Arrhenius behavior is characteristic of the rate processes connected with the glass transition region in amorphous polymers and in fact for vitrifying liquids in general.

An Arrhenius plot for a generic rate constant is based on a functional relationship of the form

$$k_{rate} = A_f e^{-\Delta E^*/RT} \tag{6.4}$$

or

$$\ln k_{rate} = \ln A_f - \Delta E^*/RT, \tag{6.5}$$

where in the modern interpretation ΔE^* is an activation energy and A_f is a frequency factor. A relaxation time behaves as the inverse of a rate constant, $\tau_0 \equiv 1/k_{rate}$, so that if *Arrhenius* behavior prevailed, the appropriate function form would be

$$-\log \tau_0 = A - B/T, \tag{6.6}$$

where A, B are constants from fitting the data. It has been found from very wide experience that a simple empirical modification of eq. (6.6) can be made that allows data such as those in Figure 6.4 to be fit quite well. Specifically, the functional form is,

$$-\log \tau_0 = A - B/(T - T_\infty), \tag{6.7}$$

where T_∞ is an additional parameter. This form is often called the Vogel–Fulcher (VF) or Vogel–Fulcher–VanTammen equation [5,6,7]. Notice that this equation implies a "catastrophe" in the rate behavior. The relaxation time becomes infinitely long at a *finite* temperature, T_∞. The curve in Figure 6.4 is a fit of this form to $-\log a_T$.

The VF formulation appears to be generally applicable to the primary transition in amorphous polymers and also applicable to many other non-polymeric glass formers. However, it is of interest to note that a number of examples have been found for non-polymers where, although curvature is found in $-\log \tau_0$ vs. $1/T$ plots, very noticeable deviations from the VF equation formulation occur [8–11].

The shift factors in Figure 6.4 utilize 30 °C as the base temperature. Useful in the context of time–temperature superposition is the modification of eq. (6.9) that parameterizes the shift factor not in terms of T_∞ but in terms of the base temperature, the latter being denoted as T_0. Thus if,

$$-\log a_T = A - B/(T - T_\infty) \tag{6.8}$$

and at $T = T_0$, the reference temperature, where the shift factor $a_{T_0} = 1$, then

$$0 = A - B/(T_0 - T_\infty).$$

Subtraction of the above two relations and rearrangement gives

$$-\log a_T = -B/(T - T_\infty) + B/(T_0 - T_\infty)$$

or

$$-\log a_T = \frac{c_1(T - T_0)}{(T - T_0 + c_2)},\qquad(6.9)$$

where $c_1 = B/(T_0 - T_\infty)$ and $c_2 = T_0 - T_\infty$. This formulation is known as the Williams–Landel–Ferry or WLF equation [12].

6.2 Dielectric relaxation

6.2.1 General behavior

The dielectric data displayed in Figures 2.11 and 2.12 for PVAc were taken in the glass transition region, at 340 K. The skewed behavior of the loss vs. log f curve and the associated complex plane plot are general features, apparently found in all amorphous polymers, of the dielectric response in this region. Figure 6.5 and Figure 6.6 compare loss data and complex plane plots for amorphous poly(ethylene terephthalate) (PET) with PVAc in the glass transition region [13]. The similarity of the shapes is apparent. In Sec. 2.5 it was also mentioned that time domain behavior that is fit reasonably well by the KWW equation is equivalent to the skewed frequency domain behavior. The results in Figure 6.5 and Figure 6.6 are based on Fourier transforms of time domain scans. The original time domain scans and KWW fits are shown in Figure 6.7. Thus HN high frequency skewed loss curves and KWW-shaped time domain relaxation are essentially equivalent and apparently general signatures of the dielectric primary relaxation region in amorphous polymers.

6.2.2 Loss maps – isothermal vs. isochronal scans

As noted, the non-Arrhenius behavior of relaxation times associated with processes in vitrifying liquids is a general phenomenon. Since relaxation processes are commonly monitored in both isothermal frequency scans and isochronal temperature scans it is useful to assess the effect on the relaxation behavior in relaxation maps. Poly(vinyl methyl ether) (PVME) is used as an example here. Figure 6.8 shows a wide temperature range isochronal scan for dielectric relaxation [14]. Figure 6.9 shows some of the same data plotted isothermally vs. frequency in the glass transition region.

The frequency of the maximum, f_{max}, in the loss peak is a convenient measure of the relaxation time–temperature location. Figure 6.10 displays log f_{max} values

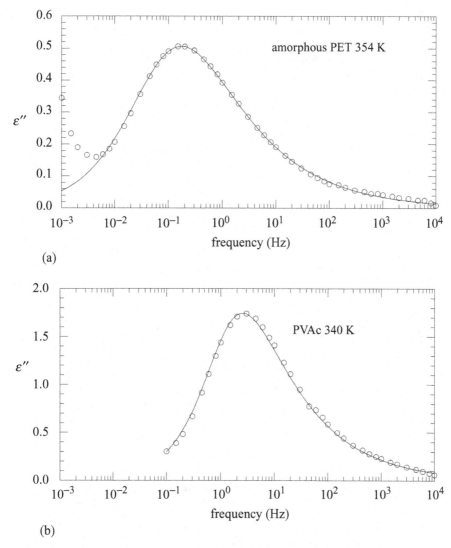

Figure 6.5 Dielectric loss in (a) PET and (b) PVAc terephthalate. Solid curves are HN function fits. Results of Boyd and Liu [13].

from Figure 6.9 vs. $1/T$. Such plots are often called *loss maps*. Also shown are values determined from the positions of the peak maxima in the isochronal scans of Figure 6.8. Although there is no requirement that the isochronal and isothermal methods give exactly the same plot, it may be seen that in this case, as is usually found, the two methods yield very similar results.

Inspection of Figure 6.8 indicates that the dielectric constant forms an envelope at higher temperature that represents the relaxed or equilibrium value. It is also

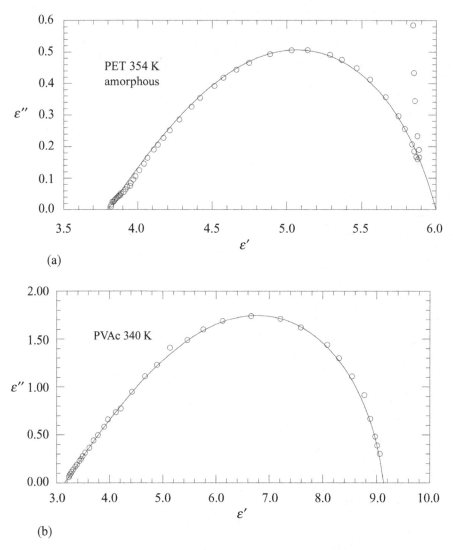

Figure 6.6 Complex plane plots for (a) PET and (b) PVAc. Solid curves are HN function fits. Same data as Figure 6.5.

apparent that this latter quantity is fairly strongly temperature dependent for the polymer at hand, PVME. There is no general rule about the magnitude or the effect of temperature on the equilibrium dielectric constant. It can be large or small and can increase or decrease (as in Figure 6.9), with increasing temperature depending on the placement of dipoles and conformational energetics within the polymer chain.

With respect to the shape in time or frequency, a noticeable temperature dependence is often found. In this context complex plane plots are instructive. For the

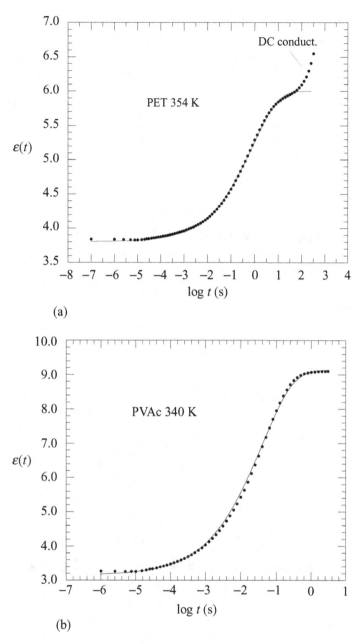

Figure 6.7 Dielectric constant measured in time domain for (a) amorphous PET and (b) PVAc. Solid curves are KWW function fits. The results in Figure 6.5 and Figure 6.6 are Fourier transforms of these data.

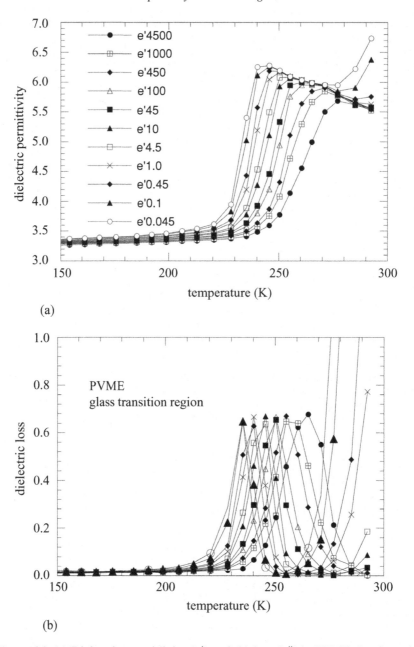

Figure 6.8 (a) Dielectric permittivity (ε') and (b) loss (ε'') in PVME. Isochronal scans at the frequencies (Hz) shown in (a). Data from Johansson and Falk [14].

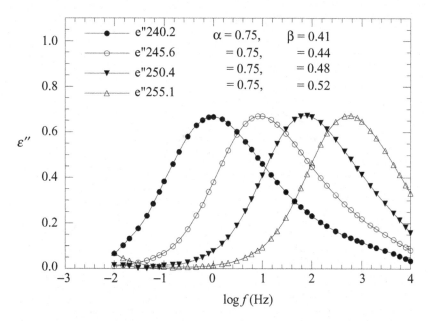

Figure 6.9 Dielectric loss in PVME, isothermal scans at the indicated temperatures (K). Curves are HN function fits, with the HN function α, β parameters shown. From same data as Figure 6.8

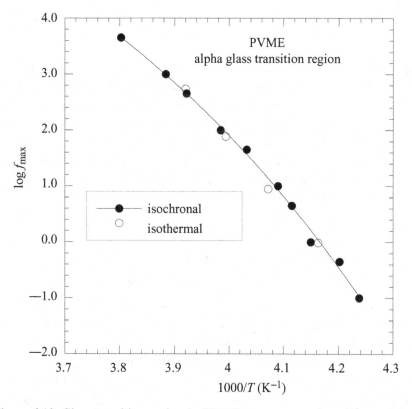

Figure 6.10 Glass transition region in PVME: temperature dependence of the frequency of loss peak maximum in both isochronal temperature scans (Figure 6.8) and isothermal frequency scans (Figure 6.8).

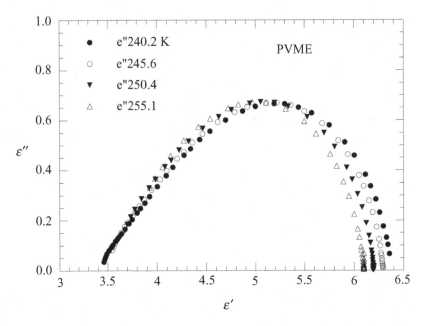

Figure 6.11 PVME: complex plane plots at four temperatures in the glass transition region.

example of PVME Figure 6.11 shows such plots at the same temperatures as in Figure 6.9. The effect of temperature in decreasing the strength of the relaxation commented on above is apparent in the low frequency zero loss intercept. The HN asymmetric skewing β parameters increase (less skewing) as temperature increases, Figure 6.9. This leads to a sharper process and the peak heights in Figure 6.9 remain nearly constant in spite of the decrease in relaxation strength alluded to above. These are not large effects but they translate into a noticeable deviation from time–temperature superposition.

6.3 Mechanical vs. dielectric relaxation

6.3.1 Comparison of scope

It is important to note that in the dielectric case essentially all of the primary relaxation region can be captured in isothermal scans vs. frequency or time. There are two reasons for this. One is that dielectric data can be experimentally measured over a broad frequency range more conveniently than in the mechanical case. More fundamental, however, is the fact that, as can be seen by comparing Figure 6.3 with Figure 6.5, the measured mechanical response covers a much broader range of $\log f$.

This is entirely the result of the greater range of change of the mechanical property. Modulus or compliance changes by orders of magnitude. The dielectric constant relaxation range is limited at the low end by the electronic polarization value in the neighborhood of 2–3 or by the presence of another relaxation process at lower temperature/higher frequency (see, e.g., Figure 6.6). Thus the detectable range of change of dielectric constant corresponds to only an order of magnitude or so of compliance change. The frequency range accompanying this is correspondingly reduced. This is illustrated in Figure 6.12, where the compliance, J', data of Figure 6.3 are plotted on a *linear* basis against log f. Figure 6.12 also shows the dielectric permittivity, ε', data for PET that are companion to the ε'' data of Figure 6.5 and also used in Figure 6.6. Both have been normalized to unit strength. In the case of J' this is accomplished by scaling the measured values by $(1/3) \times 10^{-5}$. In the case of ε', the quantity, $(\varepsilon' - 3.80)/2.20$ is plotted. When compared on this basis the dielectric relaxation and mechanical compliance behavior look much more similar. However, when both quantities are plotted on a logarithmic basis the continued measurable decrease of J' with increasing frequency is apparent.

6.3.2 Mechanical vs. dielectric relaxation location

It is of obvious interest to see to what degree the time–temperature locations of the glass transition region as determined by various spectroscopy methods compare with each other. This is especially pertinent for the mechanical and dielectric methods due to the large body of experimental studies available. Such comparison is complicated by the fact that in mechanical relaxation equilibrium values differ greatly from glassy values, i.e., by several orders of magnitude. The most direct and simple comparison would be based on measures like those invoked for dielectric relaxation. That is to say, the maximum in a loss curve is a very attractive measure on the basis of its simplicity. Mechanical log f_{max} values for maxima in J'' or E'' at various temperatures could be compared with the corresponding dielectric ones in a plot versus $1/T$. In dielectric relaxation experiments the measured dielectric constant is invariably a *retarded* function that resembles phenomenologically the mechanical compliance. However, there is no such preference in mechanical experiments. The literature is replete with examples of compliance and modulus measurements. To be sure, in dynamic frequency domain measurements there is no preference in principle between the choice of compliance and modulus expression of the results (see eq. (1.59) and eq. (1.60)). The same is true dielectrically, a *dielectric modulus* could be invoked. However, this is not often done and then it is usually to suppress the effects of ionic conduction [15,16]. Thus the question arises as to what degree it matters whether maxima in J'' or maxima in E'' are chosen for comparison with the dielectric measurements. There are two aspects

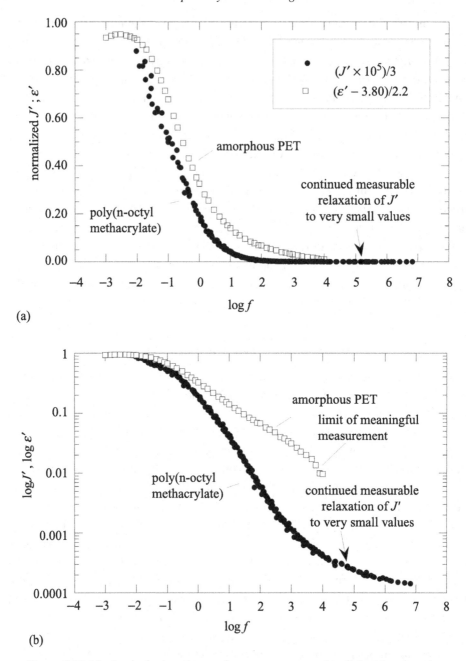

(a)

(b)

Figure 6.12 Mechanical relaxation vs. frequency compared to dielectric relaxation vs. frequency. (a) The J' data of Figure 6.3 are plotted on a *linear* basis and normalized to unity. Dielectric permittivity data for amorphous PET normalized to unity are also shown. (b) Same ordinate data as in (a) but plotted on a logarithmic basis for both.

to this question. One is purely phenomenological and deals with how much the maxima in J'' or E'' differ in location The other has to do with, on the interpretive level, which is more likely to probe the microscopic motions in a manner similar to the dielectric method. The phenomenological aspect taken up first.

Simple illustrative calculations serve to demonstrate that if the relaxed and unrelaxed moduli or compliances do not differ greatly then the choice does not matter much. However, in the case of the glass transition region, the relaxed and unrelaxed values differ by orders of magnitude. The maxima in J'' and E'' under these conditions are widely displaced from each other. This effect is shown in Figure 6.13. The same set of hypothetical data is expressed both as J^* and as E^*. In Figure 6.13(a) the glassy compliance is chosen as one-fifth of the relaxation strength, $J_r - J_u$ while in Figure 6.13(b) the glassy compliance is one-hundredth of the strength. Note that in the latter case the maximum in log J'' occurs towards the high end of the range of values for log J'. Conversely, the maximum in log E'' occurs towards the high end of the log E' values. Thus the respective log f values associated with log J'' and log E'' are more and more widely displaced as the glassy compliance becomes a smaller and smaller fraction of the relaxed value. This effect is also very apparent in Figure 1.12 vs. Figure 1.13, where the maximum in $E''(\omega)$ occurs approximately five decades higher in frequency than that for $J''(\omega)$. Use of tan δ_{max} gives a measure independent of the choice of compliance or modulus and the log f value is intermediate to those obtained from log J'' and log E''.

Having established above that the location of the loss peak maxima depends on the choice of compliance or modulus representation, the dilemma then becomes which, or neither, to chose for interpretive purposes. There is no clear answer. However, molecular considerations are helpful. According to the Rouse–Zimm theory of relaxation of flexible polymer chains the greatest contribution to the compliance comes from the longest retardation times [17]. These correspond to modes that involve the longest segment lengths, those approaching and including the entire molecular length. However, in the dielectric case, for relaxation of dipoles whose attachment is perpendicular to the chain, it must be that the effective segment length is much shorter and does not involve modes of molecular length. This suggests that the onset of softening as indicated by E''_{max} might be a better indicator for comparison with ε''_{max}. Notice also that tan δ_{max} is intermediate between the maxima in J'' and E'' and might be a reasonable choice for comparison.

A comparison of dielectric and mechanical relaxation locations for a real case is instructive. PIB is a much studied polymer mechanically. It was invoked in Chapter 1 as the illustrative example for the data manipulation for linear viscoelastic transformations between time and frequency and stress relaxation and creep. Although PIB is a seemingly non-polar hydrocarbon, substantial valence angle distortions in the skeletal carbon–carbon bonds [18,19] result in non-cancellation of whatever

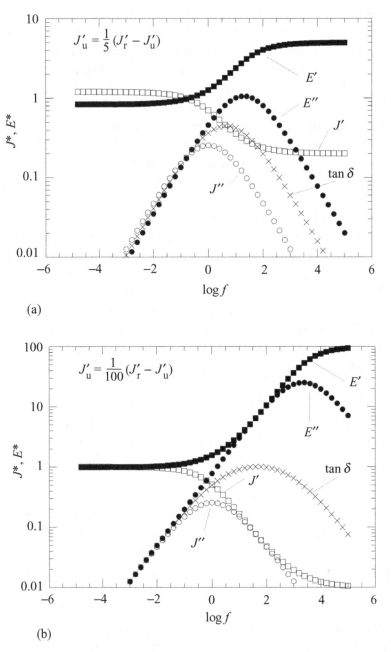

Figure 6.13 Illustration of the displacement of J'', E'' maxima from each other along the frequency axis: (a) the glassy compliance is one-fifth of the relaxation strength; (b) the glassy compliance is one-hundredth of the relaxation strength.

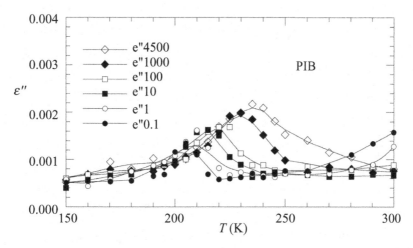

Figure 6.14 Dielectric loss in PIB. Isochronal plots vs. temperature at the indicated frequencies (Hz) [23].

dipole moment the C–H bonds possess. Thus the molecule is slightly polar and is polar enough that dielectric measurements can be effected [20–23]. Figure 6.14 shows isochronal dielectric loss vs. temperature at various frequencies for PIB.

The derived $\log f_{max}$ values from Figure 6.14 plus additional values at intermediate frequencies not displayed in Figure 6.14 are shown in a loss map in Figure 6.15. Also shown are mechanical results for PIB. They are derived from the dynamic modulus data for $E'(\omega)$, $E''(\omega)$ results in Figure 1.12, which are in turn derived from Figure 1.9. The curve in Figure 1.9 was originally constructed from stress relaxation data at various temperatures and combined into a master curve via time–temperature superposition [1]. The shift factors involved have been parameterized in terms of the WLF equation. Therefore the values of $\log \omega_{max}$ at various temperatures can be determined from $\log \omega_{max} = 6.0$ at E''_{max} in Figure 1.12 (at 25 °C the master curve reduction temperature) and the WLF parameters. The values thus determined are shown in Figure 6.15 along with the dielectric results.

It is evident that there is reasonable accord between the location of the mechanical process as represented by E''_{max} and the location of the dielectric process. In Figure 1.14, J''_{max} is seen to lie some 5.5 decades lower in frequency than E''_{max} in Figure 1.12. Thus, although the temperature variation would be the same, the correspondence in location would be very different from the dielectric result. Thus the conjecture that higher frequencies emphasized by the loss modulus E'' and initial softening correspond better to the length scales involved in dielectric relaxation in the transition region seems to be very reasonable.

Comparisons between process locations are often made on the basis of isochronal scans in which the mechanical results are limited to one or a few frequencies.

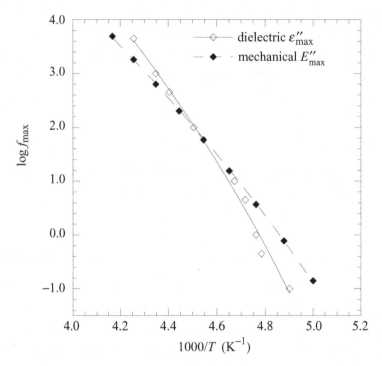

Figure 6.15 Comparison of mechanical relaxation location with dielectric location for PIB. Dielectric $\log f_{max}$ values are from Figure 6.14 along with intermediate temperature points not shown there. The solid curve is a VF equation fit, eq. (6.9), with $A = 14.5$, $B = -1127$, $T_\infty = 132$ K. Mechanical values were determined as described in the text. The mechanical points were calculated at 5 K intervals over the range of the dielectric measurements. Data of Boyd and Liu [23].

Figure 6.16 shows mechanical data from a torsion pendulum at a single frequency for the glass transition region in amorphous PET. Figure 6.17 compares the location of the glass transition region in PET from dielectric isochronal temperature scans with the mechanical G''_{max} result. The mechanical value is found to be at a frequency ~1–2 decades higher or at a temperature 4–5 K lower than its dielectric counterpart. Notice that according to Figure 6.13 the choice of $\tan \delta$ or J'' rather than G'' for Figure 6.16 would lower the frequency of the maximum isothermally and therefore increase the temperature of the maximum in an isochronal scan. This could possibly then bring closer correspondence between the mechanical and dielectric processes but also could result in overshoot. In fact the data in Figure 6.16 lead to displacement of both J''_{max} and $\tan \delta_{max}$ to higher temperature such that no maximum in either is observable in the available data range. Thus the mechanical J''_{max} and $\tan \delta_{max}$ points corresponding to the G''_{max} point do indeed overshoot and lie to an unknown extent on the other side of the dielectric data curve in Figure 6.17.

Table 6.1. *Summary of mechanical and dielectric relaxation in the primary transition region*

Descriptor	Mechanical (M)	Dielectric (D)
strength		
relaxed (r) vs. unrelaxed (u) modulus or dielectric constant)	$E_r \ll E_g$ $E_r \sim E_g / 1000$	$\varepsilon_r > \varepsilon_u$ highly variable depending on moments and densities of dipoles and their statistical correlation but: ε_r often several times larger than ε_u
temperature dependence of location	WLF VF $T_g(M) \sim T_g(D)$	WLF VF
shape vs. log t or log f	approximately linear intermediate region in log $E(t)$ vs. log t (*appears* broader than dielectric because of $E_r \ll E_u$ effect)	vs. log f: HN high frequency skewing vs. log t: KWW

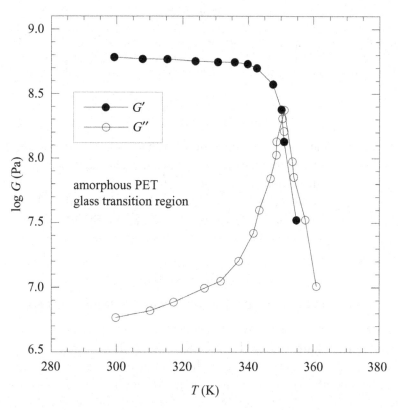

Figure 6.16 Shear modulus components, G', G'' for amorphous PET. Torsion pendulum ~1 Hz. Data of Illers and Breuer [24].

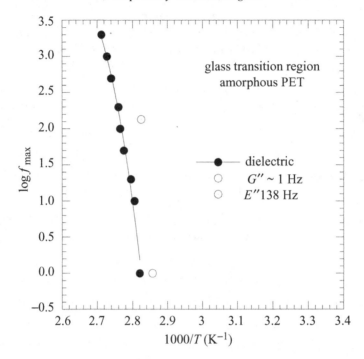

Figure 6.17 Loss map for dielectric loss peak and mechanical G'' peak. Filled circles are dielectric results [25]. Open circles are the G''_{max} mechanical result of Figure 6.16, and E''_{max} at 138 Hz (tensile vibratory forced oscillation, data of Takayanagi [26]).

In summary, in the primary transition region there is a general correspondence between the time–temperature location of the mechanical and dielectric processes. When loss maxima are compared G''_{max} may well be the best choice for comparison with dielectric data. Table 6.1 briefly summarizes and compares the salient features of mechanical and dielectric relaxation in the primary transition region.

6.4 NMR relaxation

6.4.1 NMR spin-lattice relaxation and segmental dynamics

The reorientation of ^{13}C–H or C–^{2}H vectors that gives rise to spin-lattice relaxation as measured in the ^{13}C or ^{2}H spectra, respectively (Chapter 3), is related to polymer segmental dynamics. However, the relationship between measured T_1 and NOE and the underlying motion of the polymer is not transparent. Because NMR relaxation experiments do not directly measure the $P_2(t)$ orientation autocorrelation function (eq. (3.14)), detailed information about polymer segmental dynamics from

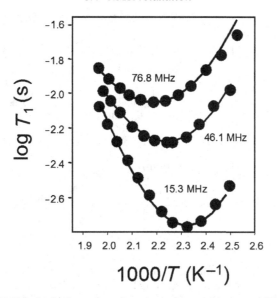

Figure 6.18 NMR spin-lattice relaxation times for perdeuterated PS as a function of inverse temperature for several resonance frequencies. The solid lines are fits using the mKWW form of the autocorrelation function given by eq. (6.10) and a VF temperature dependence. Data from He *et al.* [27].

these measurements requires construction of $P_2(t)$ based upon fitting the experimental spectra. This procedure is greatly facilitated by obtaining data over a wide range of temperatures and several resonance frequencies as shown, for example, in Figure 6.18 for a perdeuterated atactic poly(styrene) (PS) melt of approximately 2000 molecular weight [27]. In order to reproduce the measured T_1 values, $P_2(t, T)$ for $C–^2H$ vectors was determined by fitting the modified KWW (mKWW) equation

$$P_2(t, T) = a_{\text{lib}} \exp\left(-\frac{t}{\tau_{\text{lib}}}\right) + (1 - a_{\text{lib}}) \exp\left[\left(-\frac{t}{\tau_{\text{seg}}(T)}\right)^{\beta}\right] \quad (6.10)$$

to provide the best description of T_1 values through use of eq. (3.12). Here a_{lib} is the amplitude of the short-time librational contribution to the relaxation of the $C–^2H$ vectors and τ_{seg} is the segmental relaxation time, which was assumed to have a VF temperature dependence. The fit $P_2(t, T)$, shown in Figure 6.19, reproduces T_1 for PS as a function of temperature and field strength quite accurately as shown in Figure 6.18.

The representation of $P_2(t, T)$ as a sum of librational and segmental relaxation contributions (eq. (6.10)), while providing a good representation of the temperature and frequency dependence of the measured T_1 for PS melts, is not unique. In order to determine how accurately this approach to fitting experimental NMR relaxation

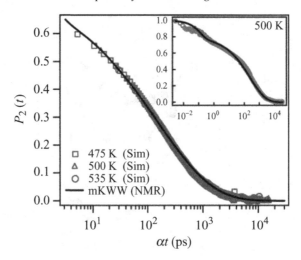

Figure 6.19 C–^{2}H $P_2(t,T)$ function for perdeuterated atactic PS determined from NMR data and MD simulations. The solid line is the mKWW autocorrelation function obtained from fits to the NMR data, shifted in time to overlap the simulation curves. $P_2(t)$ from MD simulations at 535, 500, and 475 K were superimposed using $\alpha = 1, 0.48$, and 0.17, respectively. The inset shows $P_2(t)$ at 500 K from MD simulations over a larger time window. The solid line in the inset is a mKWW fit (eq. (6.10)) with $\beta = 0.43$, $\tau_{seg} = 250$ ps, $a_{lib} = 0.195$, and $\tau_{lib} = 0.13$ ps. Taken from He *et al.* [27] with permission.

data obtained over a wide range of temperature and resonance frequency reproduces the underlying molecular motion, MD simulations of an atactic PS melt of the same molecular weight utilizing an all-atom quantum-chemistry-based potential were undertaken [27]. The $P_2(t, T)$ function for the C–^{2}H orientation obtained from simulations at three temperatures is compared with that obtained from fitting experimental T_1 in Figure 6.19. The agreement between simulation and experiment is excellent.

From the $P_2(t, T)$ fit to reproduce the experimental T_1 values, relaxation correlation times, given as

$$\tau_{seg,c} = \int_0^\infty \exp\left[\left(-\frac{t}{\tau_{seg}(T)}\right)^\beta\right] dt, \tag{6.11}$$

were determined and are shown in Figure 6.20(a). Also shown are segmental relaxation times determined from the $P_2(t, T)$ obtained from MD simulation. As with $P_2(t, T)$ itself, the segmental correlation times obtained from simulation are in excellent agreement with those extracted from experiment. Figure 6.20(b) reveals that segmental motions in PS probed by NMR relaxation measurements at high

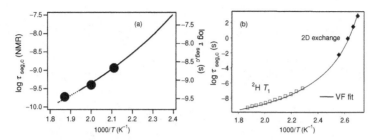

Figure 6.20 (a) Segmental correlation times of perdeuterated atactic PS from ^{2}H NMR T_1 experiments (solid lines) and MD simulations (points). The simulation correlation times have reduced by a factor of 1.8. The dashed lines represent extrapolations of the NMR results based upon a VF temperature dependence. (b) Comparison of correlation times with 2D exchange experiments obtained on high molecular weight samples from Kaufmann *et al.* [28]. Data from the relaxation measurements have been horizontally shifted to account for the molecular weight dependence of the glass transition temperature. The solid line is a VF fit to the data. Taken from He *et al.* [27] with permission.

temperature show the same VF temperature dependence as segmental relaxation times probed by 2D exchange measurements [28]. Furthermore, the VF temperature dependence of segmental relaxations probed by NMR spin-lattice relaxation is identical to that found for the chain terminal relaxation as probed by the viscosity.

Because of the excellent agreement between spin-lattice relaxation times and segmental correlation times obtained from MD simulations and experiment such as shown in Chapter 3 for PBD and above for atactic PS, it is possible to utilize the detailed molecular information provided by MD simulations to investigate the relationship between segmental dynamics as probed by NMR relaxation measurements and the underlying conformational motions that lead to the α-relaxation in polymer melts. The results of such an MD simulation study carried out for a PBD melt [29] are summarized in Figure 6.21. This PBD melt, which is discussed extensively throughout this book, consists of 40 PBD chains comprising 30 repeat units with a random 50% *trans*, 40% *cis* and 10% vinyl architecture. All MD simulations of PBD discussed in this book were performed using a quantum-chemistry-based united atom potential [30]. In Figure 6.21 the segmental relaxation time τ_{CH} (eq. (6.11)) for the C–H vector associated with sp^3 carbons of a *trans* monomer C_{sp^3}–$C_{sp^2} \overset{trans}{=} C_{sp^2}$–$C_{sp^3}$ (see also Figure 3.6) is compared with the mean waiting time between conformational transitions for *trans* allyl C_{sp^3}–C_{sp^3}–$C_{sp^2} \overset{trans}{=} C_{sp^2}$ and β (alkyl) C_{sp^2}–C_{sp^3}–C_{sp^3}–C_{sp^2} dihedrals as well as the (integrated, eq. (6.11)) relaxation times τ_{TACF} for the TACFs (eq. 5.3)) for these dihedrals. The rates of conformational transitions are well represented by an Arrhenius temperature dependence

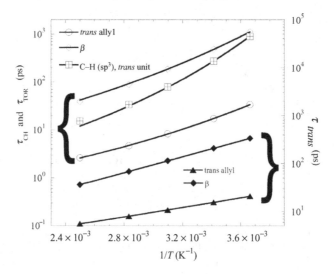

Figure 6.21 Mean conformational transition times τ_{trans}, P_2^{CH}, autocorrelation times τ_{CH}, and integrated torsional autocorrelation times τ_{TACF} for PBD from MD simulations. Solid lines are exponential (mean conformational transition times) or VF (autocorrelation times) fits to the data. Data from Smith *et al.* [29].

whose activation energy reflects the intrinsic (single-chain) conformational energy barriers in PBD. τ_{CH} exhibits non-Arrhenius temperature dependence that can be well described by a VF relation. Furthermore, τ_{CH} is much longer than the mean waiting time between transitions, indicating that many conformational transitions (on average) of each allyl and β dihedral are required to cause complete reorientation of the *trans* sp^3 C–H vectors. In contrast there is a much closer correspondence between τ_{TACF} for the *trans* allyl and β dihedrals and τ_{CH}. It can be seen in Figure 6.21 that τ_{TACF} exhibits similar temperature dependence to τ_{CH} and the segmental relaxation times as given by τ_{CH} actually lie between τ_{TACF} for the low conformational energy barrier *trans* allyl (short τ_{TACF}) and higher conformational energy barrier alkyl (long τ_{TACF}) dihedrals over the entire temperature range, indicating that reorientation of the *trans* sp^3 C–H vectors occurs as a result of a combination of *trans* allyl and β dihedral transitions. The disconnection between the rate of conformational transitions and τ_{TACF} apparent in Figure 6.21 is discussed in detail in Sec. 8.2.

6.4.2 NMR relaxation and normal mode dynamics

While C–H vector reorientation occurs primarily through local conformational dynamics, complete relaxation of the $P_2(t)$ C–H vector autocorrelation requires larger scale motions of the polymer chain, i.e., the normal (or Rouse) modes, described in Appendix AI. Indeed, in MD simulations of unentangled polymer melts [31,32], the C–H $P_2(t)$ vector autocorrelation function was found to relax mostly through fast segmental dynamics which is related to conformational dynamics as discussed above, a process expected to be largely independent of the molecular weight of the chain. However, these simulations revealed a long-time tail in the C–H vector $P_2(t)$ autocorrelation function whose relaxation time was found to be of the order of the Rouse time of the polymer chains. This second, slow, relaxation should be dependent on the molecular weight of the polymer if it is indeed related to the chain normal mode dynamics. Unfortunately, no systematic simulation study of the influence of polymer molecular weight on the behavior of the long-time tail in the C–H vector $P_2(t)$ autocorrelation function has been conducted.

In order to better understand the source and behavior of the slow dynamics observed in the relaxation of the C–H vector $P_2(t)$ autocorrelation function from MD simulations of unentangled polymer melts ^{13}C spin-lattice relaxation measurements have been conducted over a wide range of temperature and resonance frequency for melts of poly(ethylene) of two molecular weights below the entanglement molecular weight, specifically $C_{44}H_{90}$ and $C_{153}H_{308}$ [33]. For $C_{153}H_{308}$ the measured T_1 and NOE values were reproduced using the following form for $P_2(t)$:

$$P_2(t) = A_{seg} \exp\left[-\frac{t}{\tau_{seg}}\right] + (1 - A_{seg})G_{Rouse}(t), \qquad (6.12)$$

where $G_{Rouse}(t)$ is generated from the Rouse modes (Appendix AI) as described by Qiu and Ediger [33]. The first term accounts for the majority of the decay that is due to segmental motion and the second for the decay of the C–H vector, that depends on normal mode dynamics. Since measurements were performed well above T_g for poly(ethylene), both τ_{seg} and the Rouse time τ_R (eq. (AI.7)) were assumed to have an Arrhenius temperature dependence whose prefactor and activation energy, along with the amplitude A_{seg}, were adjustable parameters. The best fit to experimental T_1 and NOE values for $C_{153}H_{308}$ is shown in Figure 6.22(a). The fitted Rouse time was found to be in good agreement with values obtained for the same molecular weight poly(ethylene) from self-diffusion and viscosity measurements. This quality of fit could only be obtained when ten or more normal modes were utilized in the description of $P_2(t)$, confirming the important role of global chain dynamics C–H vector reorientation in unentangled polymer melts observed in the MD simulation studies mentioned previously. Furthermore, the dependence of the normal mode amplitudes and relaxation times on the mode index was taken from MD simulations of a $C_{100}H_{202}$ melt [31] for modes beyond the fifth mode. When Rouse scaling for the normal mode amplitudes and relaxation times (eq. (AI.5)) was used the best fit obtainable to experimental T_1 and NOE values was significantly poorer. The need

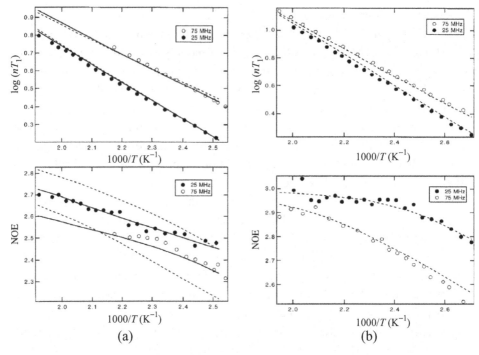

Figure 6.22 (a) Representation of T_1 and NOE values for $C_{153}H_{308}$ poly(ethylene) obtained with ten Rouse modes (solid lines) and five Rouse modes (dashed lines). (b) T_1 and NOE values for $C_{44}H_{90}$. Dashed lines are predictions of utilizing eq. (6.12) for $C_{44}H_{100}$ with the segmental relaxation parameters obtained for $C_{153}H_{308}$ and scaling the Rouse time according to the difference in molecular weight. Taken from Qiu and Ediger [33] with permission.

to utilize scaling of normal mode amplitudes and relaxation times obtained from simulations confirms the serious deviations of the chain normal modes from the Rouse model observed in simulations of $C_{100}H_{202}$ [31].

Finally, Figure 6.22(b) shows the predicted T_1 and NOE values for $C_{44}H_{90}$ obtained using the segmental relaxation parameters obtained for $C_{153}H_{308}$ and scaling the Rouse time according to the difference in molecular weight, i.e., $\tau_R(C_{44}H_{90}) = \tau_R(C_{153}H_{308})[44/153]^2$. Hence, there were no adjustable parameters in the description of experiment for $C_{44}H_{90}$. The excellent agreement obtained with experiment for T_1 and NOE using parameters obtained for the higher molecular weight melt is an important verification of the importance of slow normal-mode dynamics to the reorientation of the C–H vectors in poly(ethylene).

6.5 Neutron scattering

As introduced in Chapter 4, varying the momentum transfer q in dynamic neutron scattering experiments allows one to probe atomic motion on various length scales.

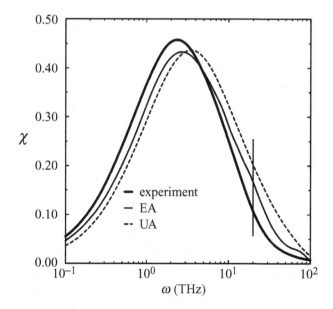

Figure 6.23 The susceptibility of a $C_{100}H_{202}$ melt at 509 K as a function of angular frequency for $q = 1.8$ Å^{-1} from incoherent neutron TOF measurements and MD simulations using explicit (EA) and united atom (UA) representations. The vertical line indicates the upper frequency limit of the experiment. Data from Smith *et al.* [35].

Hence, QENS studies of polymer melts can probe atomic motions associated with vibrations, librations, segmental motions, normal (Rouse) modes and even polymer entanglements. While much progress has been made on the latter (e.g., as reviewed in [34]), we will concentrate on the behavior of unentangled polymer chains where direct comparison between neutron scattering and MD simulations has led to a vastly improved understanding of polymer melt dynamics.

6.5.1 Librations and conformational transitions probed by dynamic neutron scattering

Figure 6.23 shows the dynamic susceptibility $\chi(q, \omega) = \omega S_{inc}(q, \omega)$ obtained for an unentangled poly(ethylene) melt ($C_{100}H_{202}$) at 509 K from both neutron scattering measurements and the MD simulations [35] shown in Figure 4.3. The main relaxation peak in the susceptibility and the decay of $S_{inc}(q, t)$ (Figure 4.3) are well reproduced by the simulation models. The vertical bars indicate the maximum frequency (energy transfer) or minimum time for the experiment at each momentum transfer. The explicit atom (EA) simulation, where all atoms are maintained in the representation of poly(ethylene), displays high frequency peaks with corresponding short-time decay of $S_{inc}(q, t)$ that are due to C–C–H and H–C–H bond angle

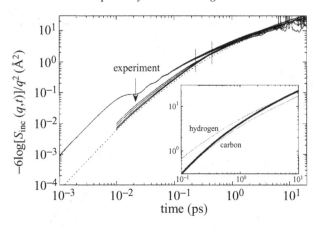

Figure 6.24 The main panel shows hydrogen mean-square displacement obtained directly from MD simulation and from application of eq. (6.13) to $S_{inc}(q,t)$ for various q values from EA simulations and neutron TOF measurements for $C_{100}H_{202}$ at 509 K. The dashed line is for carbon atoms from the EA simulations. The vertical lines show the lower time limits for the experimental data. In the inset, which shows longer timer behavior, dashed lines are for carbon and hydrogen atoms for the EA simulations, and the heavy line is the experiment for $q = 0.8$ Å^{-1}. Reproduced from Smith *et al.* [35] with permission.

vibrations. These peaks are absent in the experiment because they are outside the experimental frequency window and are absent in the united atom (UA) simulation because these degrees of freedom have been eliminated in the UA representation.

In Figure 6.24 the Gaussian approximation for the self part of the van Hove function has been used to relate the incoherent scattering function to the mean-square displacement of the hydrogen atoms [36]:

$$\langle \Delta R^2(t) \rangle = \frac{-6 \ln [S_{inc}(q, t)]}{q^2}, \tag{6.13}$$

where $\langle \Delta R^2(t) \rangle$ is the mean-square displacement of the scattering centers (hydrogen atoms) after time t. The hydrogen atom displacement shows an oscillatory contribution on these time scales due to the bond angle vibrations. The experimental proton displacement curves, given by applying eq. (6.13) to the experimental $S_{inc}(q, t)$, follow the hydrogen displacement curve from simulation at times larger than a few picoseconds. For shorter times (lying outside of the experimental time windows indicated by the vertical bars), the extrapolated displacement from experiment traces the backbone motion since the high frequency bending motions are not seen experimentally. The inset in Figure 6.24 shows the cross-over in the experimental data from tracing the hydrogen displacement to tracing the backbone displacement in more detail.

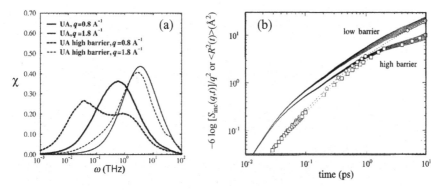

Figure 6.25 (a) The (dynamic neutron scattering) susceptibility for a $C_{100}H_{202}$ melt at 509 K as a function of angular frequency for $q = 0.8$ Å^{-1} and $q = 1.8$ Å^{-1} from MD simulations using the UA and UA high barrier models. (b) Carbon mean-square displacements (triangles are from the low barrier model, squares are from the high barrier model) and apparent hydrogen mean-square displacements (lines, from application of eq. (6.13)) as a function of time from UA simulations. Reproduced from Smith *et al.* [35] with permission.

The question remains as to the source of the molecular motions leading to the main relaxation peak in the susceptibility (Figure 6.23) and the decay of $S_{inc}(q, t)$ (Figure 4.3). The two prime candidates are librational motion in the *trans* and *gauche* minima of the backbone dihedral potentials and transitions between these minima (conformational transitions). By modifying the dihedral potential, specifically by increasing the barriers in the dihedral potential between the *trans* and *gauche* minima in poly(ethylene) by a factor of 2, it is possible to slow the rate of conformational transitions exponentially while leaving the librational frequencies of the torsions relatively uninfluenced. Figure 6.25(a) shows a comparison of the susceptibility spectra for $q = 0.8$ Å^{-1} and $q = 1.8$ Å^{-1} for the original potential and the potential with increased barriers. Based on the mean-square displacement at which $S_{inc}(q, t)$ decays to a value of 1/e, it can be seen that a momentum transfer of $q = 1.8$ Å^{-1} probes motions on a length the scale of 1–2 Å, i.e., $\langle \Delta R^2(t) \rangle = -6 \ln(1/e)/q^2 = 6/(1.8 \text{ Å}^{-1})^2 = 2$ Å^2. On this scale the spectrum shows little change when the barriers are increased, hence the main contribution to proton motion comes from librations. For $q = 0.8$ Å^{-1}, or motions on the scale of 3–4 Å, librations and transitions contribute about equally to the susceptibility and are clearly separated when the torsional barriers are increased.

The length scale dependent contribution of librational motions and conformational transitions to the motion of the polymer is further illustrated in Figure 6.25(b), where the mean-square displacements of the backbone carbon atoms for the original and modified (UA) potentials are compared. Also shown are the apparent hydrogen

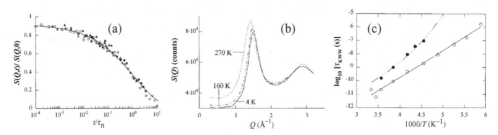

Figure 6.26 (a) Scaling representation of the NSE data for deuterated PBD at $q = 1.48$ Å^{-1} for temperatures ranging from 280 K to 220 K. The solid line is a KWW representation of the data. (b) Static structure factor for deuterated PBD. (c) Temperature dependence of the characteristic KWW times obtained from the fits of $S_{coh}(q, t)$ to stretched exponentials with $\beta = 0.41$ at $q = 1.48$ Å^{-1} (filled symbols) and 2.71 Å^{-1} (open symbols). The dashed-dotted line corresponds to the VF temperature dependence of the viscosity for PBD and the solid line to the Arrhenius temperature dependence of the dielectric β relaxation. Taken from Arbe *et al.* [37] with permission.

mean-square displacements obtained by applying eq. (6.13) to $S_{inc}(q, t)$ from the same simulations. For times less than about 1 ps the backbone atom displacements are nearly identical for the two models and the apparent hydrogen atom displacements show little splaying, indicative of Gaussian-distributed displacements. These observations, combined with the fact that the hydrogen displacements at 1 ps correspond well with the length scale associated with $q = 1.8$ Å^{-1}, support the conclusion that polymer motion on time scales less than 1 ps and length scales of around 1 Å or less is due primarily to torsional librations. The diverging of the backbone atom displacements for the two models and the splaying of the apparent hydrogen displacements point to the increasing importance of conformational transitions, which lead intrinsically to heterogeneous (non-Gaussian) atom displacements, at times longer than 1 ps and length scales greater than 1 Å.

6.5.2 Structural relaxation and dynamic neutron scattering

Neutron scattering has played a central role in establishing the fundamental relationship between structural relaxation in polymer melts, microscopic measurements of dynamics such as provided by dielectric and NMR spectroscopy and macroscopic probes of polymer motion, particularly rheological measurements. Figure 6.26(a) shows the intermediate coherent structure factor for a fully deuterated PBD melt obtained on the NSE spectrometer IN11 at the Institut Laue-Langevin at a momentum transfer of $q = 1.48$ Å^{-1} [37]. As can be seen in Figure 6.26(b), this momentum transfer corresponds to the first maximum in the static structure factor which reflects largely intermolecular near-neighbor correlations. Data obtained over a wide range of temperatures were reduced to the single master curve shown in Figure 6.26(a) using the KWW relationship

$$\frac{S_{\text{coh}}(q, t)}{S_{\text{coh}}(q, 0)} = A \exp\left[-\left(\frac{t}{\tau_{\text{KWW}}(T)}\right)^{\beta}\right] \tag{6.14}$$

with a temperature dependent τ_{KWW} shown in Figure 6.26(c).

The decay of $S_{\text{coh}}(q, t)$ at the first peak in the static structure factor reflects the rate at which nearest-neighbor interchain correlations relax in the polymer melt, a process referred to as structural relaxation. The stretching exponent ($\beta = 0.41$) used in eq. (6.14) to create the master curve shown in Figure 6.26(a) was taken from dielectric measurements. As discussed in Chapter 2, dielectric relaxation is sensitive to local conformational motions. Hence, structural relaxation as measured by dynamic neutron scattering at the first peak in the static structure factor and dielectric relaxation exhibit the same non-Debye (i.e., KWW) relaxation behavior and are sensitive to motions on the length scale of a polymer segment, i.e., $2\pi/1.48 \text{ Å}^{-1} \approx$ 4 Å for dynamic coherent neutron scattering and several dihedrals for dielectric relaxation (Sec. 6.3). The VF temperature dependence of the structural relaxation times τ_{KWW}, illustrated in Figure 6.26(c), is well described by the temperature dependent viscosity, indicating that structural/segmental relaxation as probed by coherent neutron scattering is sensitive to the same motions that lead to large scale relaxation of the polymer chains.

6.5.3 Neutron scattering and Rouse dynamics

By probing mixtures of deuterated and protonated polymer chains, NSE techniques allow measurement of the intermediate coherent dynamic structure factor for a single polymer chain on length scales much larger than the radius of gyration (R_g) of unentangled chains down to length scales on the order of R_g for these chains on a time scale ranging from 50 ps up to 100 ns. The normalized intermediate single-chain dynamic structure factor $S'(q, t)$, given by

$$S'(q, t) = \frac{\left\langle \sum_{j',k'}^{N'} \exp i\mathbf{q} \cdot (r_{k'}(t) - \mathbf{r}_{j'}(0)) \right\rangle}{\left\langle \sum_{j',k'}^{N'} \exp i\mathbf{q} \cdot (r_{k'}(0) - \mathbf{r}_{j'}(0)) \right\rangle}, \tag{6.15}$$

where a "prime" indicates summation over atoms belonging to the same chain only, is shown in Figure 6.27(a) for an unentangled PBD melt at 353 K as measured on the NSE spectrometer at the FRJ-2 reactor in Jülich [38]. Also shown is $S'(q, t)$ from MD simulations of the same polymer (microstructure, molecular weight and temperature, see Sec. 6.4). Excellent agreement between experiment

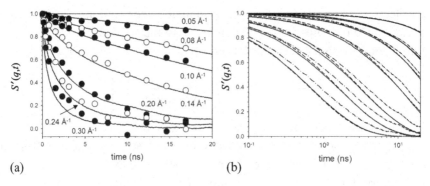

Figure 6.27 (a) The single-chain coherent intermediate scattering function for PBD at 353 K from experiment (symbols) and MD simulation (lines). The time scale for experiment has been scaled by 0.77 to account for differences in center-of-mass diffusion between simulation and experiment. (b) Single-chain coherent intermediate scattering function for PBD at 353 K from simulation (dash-dot lines) compared to the prediction of the Rouse model (dashed lines) and the scattering from simulation calculated using the dynamic Gaussian assumption for the monomer displacements (solid lines). Momentum transfers are same as in (a). Data are taken from Smith *et al.* [38].

and simulation can be seen at all times for all q values investigated. However, a comparison of $S'(q, t)$ from simulation and that predicted by the Rouse model (eq. (AI.20)), shown in Figure 6.27(b), reveals that the Rouse model does a poor job in reproducing the single-chain dynamic structure factor, especially for larger q values. In particular, the Rouse model predicts too fast a decay of $S'(q, t)$ for larger q values. This behavior has been observed for a number of polymer melts (e.g., as reviewed by Paul and Smith [34]). Since the Rouse model forms the basis of our understanding of polymer chain dynamics for length scales larger than the segmental length scale, understanding the source of these discrepancies is central to understanding polymer melt dynamics. For PBD the normal modes as obtained from simulation (eq. (AI.4)) were analyzed in detail and it was found that the Rouse model does a good job in describing the amplitudes and relaxation times for chain normal modes contributing to $S'(q, t)$ on the q scales investigated [38]. Hence, the failure of the Rouse model to accurately reproduce $S'(q, t)$ from simulation and experiment does not lie primarily in the predicted mode amplitudes or relaxation times.

Two main approaches taken to account for the discrepancies seen between experiment and the Rouse model for $S'(q, t)$ involve increasing the mode relaxation time for large index modes compared to Rouse predictions (eq. (AI.5)), thereby slowing the relaxation of $S'(q, t)$ for large q values compared to Rouse predictions. The semi-flexible chain model (SFCM) [39] argues that the deviations occur for momentum transfers probing distances where the bending rigidity of the chains cannot be neglected, and where the large-scale random coil structure cannot yet be observed. In both the Rouse model and SFCM, the frictional drag force is assumed to be independent of wavelength. For the SFCM this leads to a mode relaxation time of

$$\tau_p = \frac{\tau_R}{p^2 + \alpha_b p^4} \qquad (6.16)$$

where the parameter α_b depends on the ratio of the Kuhn length to the contour length of the chain, yielding increased slowing of mode relaxation compared to Rouse predictions with increasing mode index p. Leaving α_b as a free fit parameter an improved fit of experimental $s_{coh}(q, t)$ with the SFCM compared to the Rouse model has been found possible but requires unreasonable values of the Kuhn length or contour length of the polymer chain. When these are fixed from structural information the SFCM has yielded only slight improvement on the Rouse model for the range of momentum transfers investigated, both for simulations [40,41] and experiments [42].

In the internal viscosity model (IVM) [43] a stiffening of the chains on local scales as well as an additional local dissipation process is introduced. The local stiffening introduces a stronger decrease of the mode relaxation times with increasing mode number than in the Rouse model, similar to the SFCM. The additional local friction is characterized by a time constant τ_0. The additional relaxation mode in conjunction with the inclusion of local stiffness allows the model to be fitted well to experimental $S'(q, t)$ [42,44]. By comparing two polymers of about equal local stiffness, PIB and polydimethylsiloxane (PDMS) [44], it was shown that PDMS, which has virtually no dihedral barriers for local rotation, can be rather well described by the Rouse model, whereas PIB, which has high rotation barriers, can only be fitted with the IVM. It was concluded that the additional process with time constant τ_0 captures the dissipation due to conformational transitions. In contrast, in a simulation of PBD and a freely rotating chain version (FR-PBD) of the same model, where all torsion potentials had been set to zero, the height of the rotation barrier had no influence on the applicability of the Rouse model [45]. Both model chains have the same static behavior at all length scales and the scattering functions for both models deviate to the same degree from the Rouse prediction. Apparently the decay of $S'(q, t)$ seen at larger q values for many polymers corresponding to slower mode relaxation for large index modes than predicted by the Rouse model is not connected with dihedral barriers as assumed by the IVM.

Based on the above discussion, it is apparent that attempts to improve agreement between predicted and experimental $S'(q, t)$ by slowing the local dynamics of the polymer are fundamentally flawed. This conclusion is further supported by the observation that the Rouse model in fact provides a good description of normal mode dynamics of the PBD melt, i.e., no indication of significant slowing of normal mode dynamics compared to Rouse predictions is seen for large index modes. Examination of the effective medium models (Rouse, SFCM, IVM) used to predict $S'(q, t)$ reveals that the predicted $S'(q, t)$ is based on the assumption that all atom displacements are Gaussian distributed (see Appendix AI), yielding the relationship between $s_{coh}(q, t)$ and atom mean-square displacements given by eq. (AI.20). The dynamic Gaussian approximation can also be utilized to predict $S'(q, t)$ from simulation, i.e., from eq. (AI.20) where the atom mean-square displacements are taken from simulations, as shown in Figure 6.27(b). In comparing $S'(q, t)$ yielded by the dynamic Gaussian assumption with the Rouse model prediction, it can be seen that the Rouse model does an excellent job in predicting this scattering function, i.e., in predicting the atom mean-square displacements obtained from simulation. Hence, the deviation between the actual $S'(q, t)$ and that predicted by the Rouse model is due to non-Gaussian behavior of the atom displacements. Non-Gaussian effects slow the decay of $S'(q, t)$ relative to that obtained for a Gaussian distribution with the same mean-square displacements, as shown in Figure 6.27(b), consistent with the observed relationship between simulation and Rouse predictions. Non-Gaussian displacements, which may be strongly influenced by intermolecular correlations that also lead to subdiffusive chain center-of-mass displacement [34,46], are not included in the single-chain effective medium type theories (Rouse, SFCM, IVM) which have been employed to analyze the dynamics in amorphous polymers.

References

[1] J. D. Ferry, *Viscoelastic Properties of Polymers*, second edn (New York: Wiley, 1970). Tabulated data in Append. D-2.

[2] D. J. Plazek, *J. Phys. Chem.*, **69**, 3480 (1965).

[3] D. J. Plazek, I.-C. Chay, K. L. Ngai, and C. M. Roland, *Macromolecules*, **28**, 6432 (1996).

[4] L. I. Palade, V. Verney, and P. Attane, *Macromolecules*, **28**, 7051 (1995).

[5] H. Vogel, *Physik Z.*, **22**, 645 (1921).

[6] G. S. Fulcher, *J. Am. Ceram. Soc.*, **8**, 339 (1925).

[7] G. van Tammen and G. Hesse, *Z. Anorg. Allg. Chem.*, **156**, 245 (1926).

[8] F. Stickel, E. W. Fischer, and R. Richert, *J. Chem. Phys.*, **102**, 6251 (1995).

[9] F. Stickel, E. W. Fischer, and R. Richert, *J. Chem. Phys.*, **104**, 2043 (1996).

[10] C. Hansen, F. Stickel, T. Berger, R. Richert, and E. W. Fischer, *J. Chem. Phys.*, **107**, 1086 (1997).

[11] C. Hansen, F. Stickel, R. Richert, and E. W. Fischer, *J. Chem. Phys.*, **108**, 6408 (1998).

[12] M. L. Williams, R. F. Landel, and J. D. Ferry, *J. Am. Chem. Soc.*, **77**, 3701 (1955).

[13] R. H. Boyd and F. Liu in *Dielectric Spectroscopy of Polymeric Materials*, edited by J. P. Runt and J. J. Fitzgerald (Washington, DC: American Chemical Society, 1997), Chapter 4.

[14] O. Johansson and T. Falk, Master of Science Thesis, Royal Institute of Technology, Stockholm (1993).

[15] H. W. Starkweather and P. Avakian, *J. Polym. Sci., Part B: Polym. Phys.*, **30**, 637 (1992).

[16] K. Mohomed, T. G. Gerasimov, F. Moussy, and J. P. Harmon, *Polymer*, **46**, 3847 (2005).

[17] J. D. Ferry, *Viscoelastic Properties of Polymers*, second edn (New York: Wiley, 1970), Chapter 9.

[18] H. Tadokoro, *Structure of Crystalline Polymers* (New York: Wiley, 1979).

[19] R. H. Boyd and S. M. Breitling, *Macromolecules*, **5**, 1 (1972).

[20] N. G. McCrum, B. E. Read, and G. Williams, *Anelastic and Dielectric Effects in Polymeric Solids* (New York: Wiley, 1967; Dover, 1991).

[21] B. Stoll, W. Pechhold, and S. Blasenbrey, *Kolloid Z.*, **250**, 111 (1972).

[22] D. Richter, A. Arbe, J. Colmenero, M. Monkenbusch, B. Farago, and R. Faust, *Macromolecules*, **31**, 1133 (1998).

[23] R. H. Boyd and F. Liu previously unpublished work.

[24] K. H. Illers and H. Breuer, *J. Colloid Sci.*, **18**, 1 (1963).

[25] J. C. Coburn and R. H. Boyd, *Macromolecules*, **19**, 2238 (1986).

[26] H. Takayanagi, from data in McCrum, Read and Williams [Ref. 20].

[27] Y. He, T. Lutz, M. D. Ediger, C. Ayyagari, D. Bedrov, and G. D. Smith, *Macromolecules*, **37**, 5032 (2004).

[28] S. Kaufmann, S. Wefing, D. Schaefer, and H. W. Spiess, *J. Chem. Phys.*, **93**, 197 (1990).

[29] G. D. Smith, O. Borodin, D. Bedrov, W. Paul, X. H. Qiu, and M. D. Ediger, *Macromolecules*, **34**, 5192 (2001).

[30] G. D. Smith, W. Paul, M. Monkenbusch, L. Willner, D. Richter, X. H. Qiu, and M. D. Ediger, *Macromolecules*, **32**, 8857 (1999).

[31] W. Paul, D. Y. Yoon, and G. D. Smith, *J. Chem. Phys.*, **103**, 1702 (1995).

[32] W. Paul, G. D. Smith, and D. Y. Yoon, *Macromolecules*, **30**, 7772 (1997).

[33] X. H. Qiu and M. D. Ediger, *Macromolecules*, **33**, 490 (2000).

[34] W. Paul and G. D. Smith, *Rep. Prog. Phys.*, **67**, 1117 (2004).

[35] G. D. Smith, W. Paul, D. Y. Yoon, *et al.*, *J. Chem. Phys.*, **107**, 4751 (1997).

[36] D. A. McQuarrie, *Statistical Mechanics* (New York: Harper and Row, 1976).

[37] A. Arbe, D. Richter, J. Colmenero, and B. Farago, *Phys. Rev. E*, **54**, 3853 (1996).

[38] G. D. Smith, W. Paul, M. Monkenbusch, and D. Richter, *Chem. Phys.*, **261**, 61 (2000).

[39] R. G. Winkler, L. Harnau, and P. Reineker, *Macromol. Symp.*, **81**, 91 (1994).

[40] G. D. Smith, W. Paul, M. Monkenbusch, and D. Richter, *J. Chem. Phys.*, **114**, 4285 (2001).

[41] V. A. Harmandaris, V. G. Mavrantzas, D. N. Theodorou, *et al.*, *Macromolecules*, **36**, 1376 (2003).

[42] D. Richter, M. Monkenbusch, J. Allgeier, *et al.*, *J. Chem. Phys.*, **111**, 6107 (1999).

[43] G. Allegra and F. J. Ganazzoli, *J. Chem. Phys.*, **74**, 1310 (1981).

[44] A. Arbe, M. Monkenbusch, J. Stellbrink, *et al.*, *Macromolecules*, **34**, 1281 (2001).

[45] W. Paul, S. Krushev, and G. D. Smith, *Macromolecules*, **35**, 4198 (2002).

[46] M. Guenza, *J. Chem. Phys.*, **119**, 7568 (2003).

7

Secondary (subglass) relaxations

7.1 Occurrence of mechanical and dielectric secondary processes

Figure 7.1 shows torsion pendulum data for PET at temperatures much lower than the data displayed in Figure 6.16 for the glass transition region. It is evident that a significant relaxation process exists. Figure 7.2 shows dielectric loss results in PET that are complementary to the mechanical results in Figure 7.1 in terms of exploring the same relaxation region. The secondary (β) dielectric process found here has a *strength* that is less than that of the glass transition region but nevertheless is quite appreciable. Determination of the strength from Cole–Cole plots using dielectric permittivity data complementary to the loss data in Figure 7.2 gives a result of $\Delta\varepsilon(\beta) = 0.5$. This is to be compared with the value $\Delta\varepsilon(\alpha) = 2.2$ from the results of Figure 6.6 for the primary transition. Comparison of the apparent *width* of the process with that of the glass transition process (cf. Figure 7.2 with Figure 6.5) shows the subglass process to be extraordinarily broad. That is, in the seven decades of frequency scan in Figure 7.2 only part of the subglass process is captured. In contrast in Figure 6.5 the entire glass transition region process is recorded over this frequency range. The very great breadth of secondary processes leads to small loss peak heights even though the strengths may be quite appreciable. It is also usual for β subglass processes to narrow with increasing temperature, as is evident in the dielectric loss data of Figure 7.2 as increasing loss peak heights and in the temperature dependence of the Cole–Cole width parameters of the fits.

In Figure 7.3 a loss map is shown that displays both the α glass transition region and the subglass β-process. The α-process shows the VF (eq. (6.7)) temperature dependence typical of primary transitions. The β relaxation is Arrhenius in form, eq. (6.6). The slope of the $1/T$ plot and the derived activation energy of the β-process are very modest compared to that of the α glass transition region. Arrhenius temperature dependence and modest activation energies are typical of secondary relaxations.

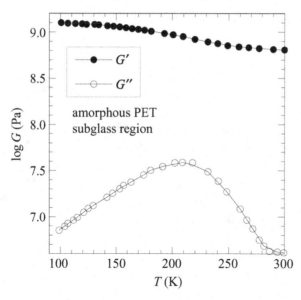

Figure 7.1 Subglass β-relaxation in amorphous PET: complex shear modulus vs. temperature from torsion pendulum experiments at $\sim$1 Hz. Data of Illers and Breuer [1].

7.2 Complexity and multiplicity of secondary processes

It is of interest to examine the frequency dispersion of secondary processes to see if they conform to simple phenomenological forms such as the HN or KWW equations that are associated with the primary transition region. Because of the great breadth of the relaxations in the frequency plane this has been a difficult judgment to make. Often the Cole–Cole equation has been invoked, as in the plots in Figure 7.2. However, the advent of wider band dielectric spectroscopy has allowed investigation of this question. The result has been the revelation that a given secondary relaxation often shows significant complexity in its structure.

Taking the above dielectric β-relaxation in PET as an example, loss vs. frequency data from Figure 7.2 at one temperature, 203 K, are shown along with a Cole–Cole plot of loss vs. permittivity in Figure 7.4. It is apparent that there are significant deviations in the data from Cole–Cole equation fits. These deviations are systematic and reproducible, appearing also at the other temperatures in Figure 7.2. The data contain inflection points that are not addressable by a simple function such as the HN equation. The inflection points suggest the presence of several poorly resolved overlapping processes.

Simulations indicate why this overlapping of processes could occur. The residual motion in the glass must be associated with motions arising from conformational transitions in the skeletal bonds of the main chain as these are the only bonds having

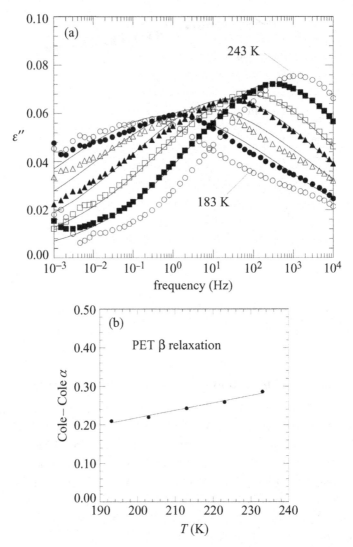

Figure 7.2 (a) Dielectric loss in the subglass β-relaxation region for amorphous PET. Data are at 10 K intervals. The curves are Cole–Cole function fits. (b) Temperature dependence of the Cole–Cole α width parameters for the fits of the PET data. Data of Boyd and Liu [2].

the requisite flexibility to drive dipolar rotational reorientation. Specifically, these are the O–C bond connecting the ester group to the ethylenic linkage, the C–C bond in the ethylenic linkage, and the CA–CD aromatic C to trigonal C bond connecting the ester group to the aromatic ring. These bonds have rotational barriers that differ significantly and increase in the order: O–C, C–C and CA–CD. MD simulations that monitor the frequency of conformational transitions show the rates decrease in the above order (Figure 7.5) [4].

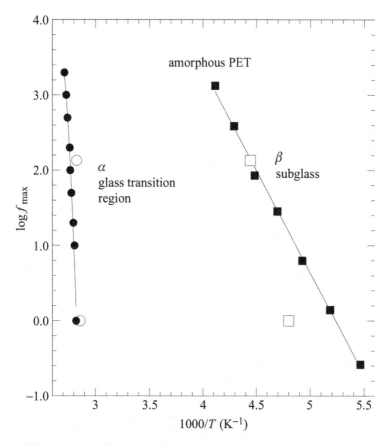

Figure 7.3 Loss map showing both the glass transition region and the subglass region in PET. Filled circles are dielectric results, open circles are mechanical log f_{max} values in the glass transition region from Figure 6.17, filled squares are dielectric log f_{max} values from Figure 7.2, and open squares are mechanical $G'_{max} \sim 1$ Hz results from Figure 7.1 and E''_{max} at 128 Hz tensile vibratory forced oscillation [3].

The considerable disparity in rates of conformational transition among the three bond types suggests that the structure in the results in Figure 7.2 and Figure 7.4 could arise from the differing rates [5]. In Figure 7.6 the results of utilizing three additive Cole–Cole processes are shown. Invoking additivity of the processes implies that they are independent though this may only be an approximation. However, it is apparent that a much better fit is obtained. The experimental activation energies deduced from the parameterization of the collective data of all the temperatures in Figure 7.2 increase in same order as in Figure 7.5. Thus it is a reasonable presumption to conclude that process "1" in Figure 7.6 is associated predominately with the C–O bond, process "2" with the C–C bond and process "3" with the CA–CD bond.

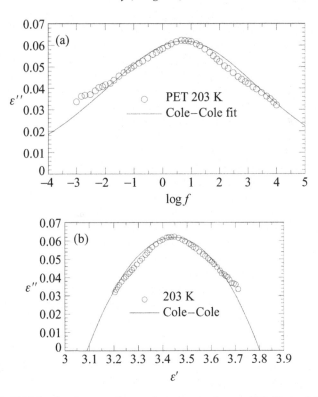

Figure 7.4 PET in the β secondary relaxation region at 203 K: in (a) the solid curve is a Cole–Cole function fit to the loss data; in (b) the solid curve is the same fit to a Cole–Cole complex-plane plot. Data of Boyd and Liu [2].

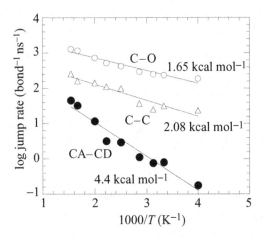

Figure 7.5 The rates of conformational transitions in PET from MD simulation at the ethylenic to ester group C–O bond, the ethylenic C–C bond, and the aromatic ring to ester group trigonal carbon CA–CD bond as a function of reciprocal temperature. Activation energies for each bond type are shown. Results of S. U. Boyd and R. H. Boyd [4].

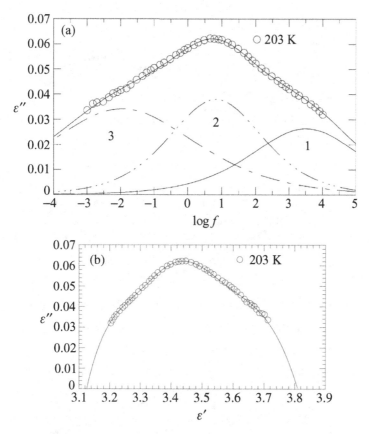

Figure 7.6 PET, three-process interpretation: (a) a fit to the loss data in Fig. 7.4 that invokes three additive Cole–Cole processes, labeled 1, 2, 3; (b) a complex plane plot with the same Cole–Cole function fit. Results of Bravard and Boyd [5].

Further insight into the nature of the complexity in the frequency spectrum of dielectric loss in secondary relaxations may be gained from results on poly(ethylene 2,6-naphthalene dicarboxylate) (PEN) with PET [5]. Both polymers share the same dynamically flexible bonds and differ only in the size of the aromatic moeity. In PEN, *two* subglass processes that are resolved in the frequency plane are observed. The higher frequency, lower temperature one is similar to the β secondary relaxation region in PET and the lower frequency, higher temperature one has been designated the β^* process [6]. Figure 7.7 shows the loss spectrum in these regions. Figure 7.8 displays Cole–Cole diagrams for PEN in the β region.

As for PET, the loss spectrum and the Argand diagrams in the β region show features that are not accommodated by the phenomenological functions in that there is an inflection in loss curves towards higher frequency and upward curvature in Argand diagrams on the low permittivity side. Unlike for PET, however, the

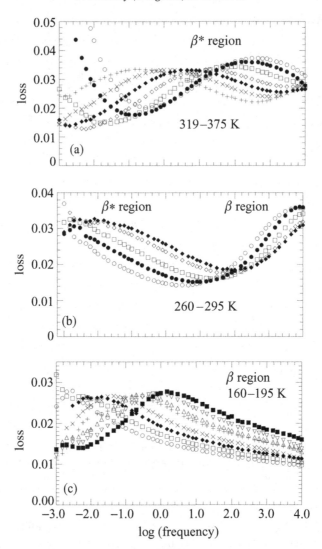

Figure 7.7 Dielectric loss in PEN in the β^*-relaxation region ((a), 319 K, 330–370 K at 10 K intervals, 375 K), in the intermediate region ((b), 260–290 K at 10 K intervals, 295 K), and in the β region ((c), 160–195 K at 5 K intervals). Results of Bravard and Boyd [5].

invocation of just two additive Cole–Cole processes gives a good fit to the data as may be seen in Figure 7.9. In the β^* region, when calculated contributions from the β-process arrived at from temperature extrapolated fitting parameters are included, a single Cole–Cole process suffices to represent the data as shown in Figure 7.10.

The conclusion drawn from the results in Figures 7.6–7.10 is that the complex nature of the secondary processes is very similar in PET and PEN. Both polymers

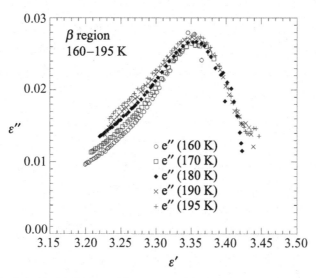

Figure 7.8 Cole–Cole diagrams for PEN in the β-relaxation region. Results of Bravard and Boyd [5].

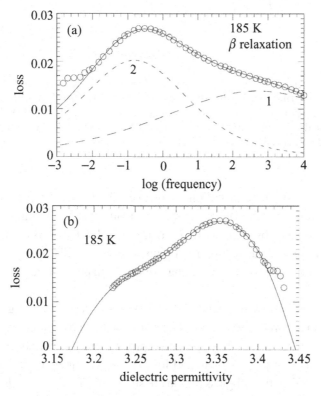

Figure 7.9 Two Cole–Cole processes fit to the PEN β-relaxation process at 185 K: (a) the fit to the loss frequency spectrum and the contributions of processes 1 and 2; (b) the Argand diagram. Results of Bravard and Boyd [5].

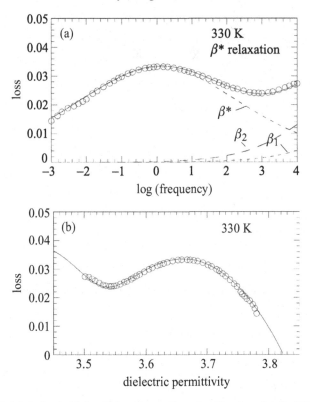

Figure 7.10 (a) A single Cole–Cole process fit to the β^* relaxation in PEN at 330 K along with contributions β_1, β_2 from the higher frequency β-relaxation constructed from temperature extrapolation of parameters in fits as in Figure 7.9. The curve through the data is the sum of the three. (b) The fit to the Argand diagram. Results of Bravard and Boyd [5].

show evidence of three contributions. However, the lowest frequency component in PEN is shifted to much lower frequency isothermally or higher temperature isochronally and is observed as a frequency resolved process. This in turn is attributed to a slower rate of reorientation at the CA–CD aromatic ring to trigonal ester carbon due to the bulkier nature of the naphthylenic unit compared to phenyl.

Another example of wider band dielectric spectroscopy revealing details of the shape of secondary processes is PVC. Historically, this polymer is probably the first one to have revealed a secondary relaxation process and this was accomplished via dielectric spectroscopy [7]. Broadband loss vs. log (frequency) curves and Cole–Cole plots for PVC in the subglass region are shown in Figure 7.11 [8]. It is apparent in Figure 7.11 that the subglass β loss process is skewed toward low frequency. That is, the magnitude of the slope of a loss curve at a given temperature is lower

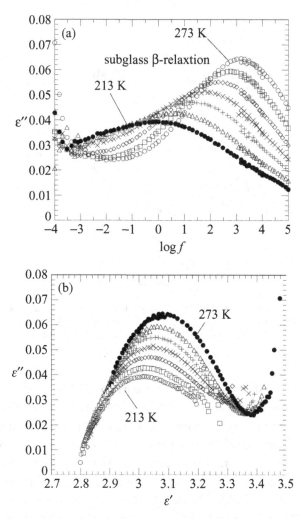

Figure 7.11 (a) Dielectric loss in PVC vs. log frequency at 10 K intervals between 213 K and 273 K. (b) Cole–Cole plot of the data. Data of J. Larsson [8].

below the frequency of maximum loss than above. Similarly, the Cole–Cole plots have higher slope magnitudes on the high frequency (low ε') side than on the low frequency (high ε') side. This is in contrast to the situation in the primary relaxation region where high frequency skewing is observed. See, for example, Figure 6.6.

7.3 Flexible side group motion as a source of secondary relaxation

There are numerous examples of subglass processes that are largely associated with motion of flexible side groups rather than motion within the chain backbone. As an

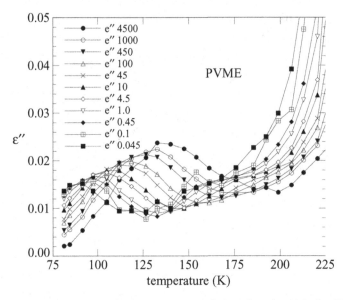

Figure 7.12 Dielectric loss data (on an expanded ordinate scale) for PVME at low temperature showing a resolved subglass process and an apparent process intermediate to the primary transition. Data of Johansson and Falk [9].

example, Figure 7.12 shows dielectric loss data for PVME in the subglass region that is companion to the loss data in Figure 6.8 in the glass transition region. This low temperature process would appear to be due to reorientation of the methyl ether groups about the attaching C–O ether bond. Figure 7.13 shows a loss map that includes both the glass transition region and the subglass process.

It has been possible to utilize molecular modeling methods in the rationalization of subglass glass relaxational behavior due to side group motion. The secondary relaxations in PVAc and poly(methyl acrylate) (PMA) are an example. These form an interesting pair because they differ only in the order of the C–O bond connection in the side group; in PVAc the ether oxygen is connected to the main chain and in PMA the carbonyl group connects to it. As seen in Figure 7.14, the locations of the secondary processes in frequency at the same temperature are similar in the PVAc and PMA polymers. However, the strengths of the secondary processes are very different in the two cases, the β-relaxation being much stronger in PMA polymers compared to PVAc polymers [10,11,12].

Application of molecular modeling involves the assumption that the main chain conformations are frozen-in below the glass transition temperature but that side group conformations continue to be dictated by equilibrium conditions [15]. Thus a population of chains is generated by Monte Carlo methods reflecting both the conformations at the glass temperature and the stereo dyads characteristic of the

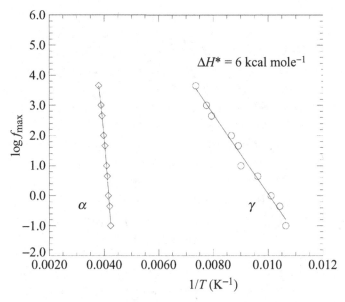

Figure 7.13 Loss map for PVME showing both the primary relaxation and the resolved subglass process. Data of Johansson and Falk [9].

polymerization process. This population is then used to calculate, at lower temperatures characteristic of the subglass relaxation, the mean-square moment of the assembly of chains with the main chain conformations fixed. In order to carry out the mean-square moment calculation statistical weights are assigned to the side group conformations. These in turn are derived from fitting the total conformational energies of a number of conformers of model compounds via force-field-based molecular mechanics methods [16]. Aside from being able to successfully represent the conformational energies of the database, including interactions between nearby ester groups, the central feature of the conformational energy calculations is that there are three stable positions for the orientation of an ester side group. These are illustrated in Figure 7.15, where a methyl acrylate type ester group attached to an all *trans* hexane molecule is shown. The flexible attaching bond for the ester group is indicated by the arrow. Under the convention that the attaching bond conformational angle is based on the atom sequence C(3)–C(4)–CD–O the stable position in the left hand structure is characterized by a value of ~60°. In the right hand structure the torsional angle is ~−150° relative to the ~60° position. Another position of +150° is equivalent.

With respect to the stable minima positions, the situation in the vinyl acetate attachment is very similar, three positions with values 60°, and ±150° relative to 60° obtain. However, the energetics are significantly different. An energy diagram from a molecular mechanics calculation representing a complete rotation about

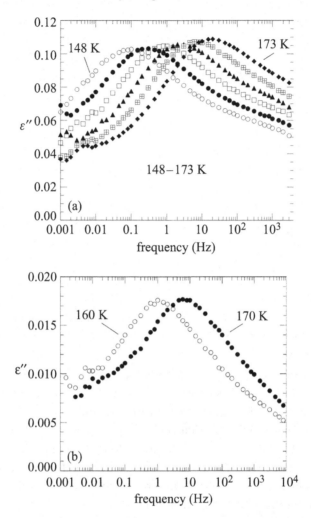

Figure 7.14 Comparison of the subglass β relaxations in (a) PMA and (b) PVAc. The frequency region of occurrence is similar but the relaxation strength is much greater in PMA. Note the difference in the two ordinate scales. PVAc data from Devereaux [13] and PMA data from Boyd and Liu [14].

the attaching bond for both the methyl acrylate and vinyl acetate cases is shown in Figure 7.16. Although the energy barriers are similar, the energy difference between the 60° and ±150° positions is greater in the vinyl acrylate case than for methyl acrylate. This increased energy difference between vinyl acetate and methyl acrylate tends to hold generally in various chain conformations and is incorporated in the statistical mechanical averaging procedure via the statistical weight parameters from the database fitting. It thus leads through Boltzmann weighting to a decrease in strength of the subglass process in vinyl acetate vs. methyl acrylate. If, for

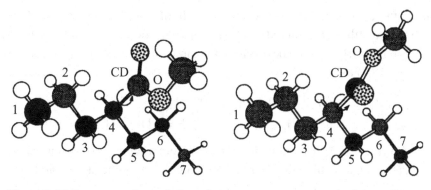

Figure 7.15 Structures of a methyl acrylate type ester group attached to an all *trans* hexane molecule. There are three stable orientations of the side group. Based on the atom sequence C(3)–C(4)– CD–O the left hand structure has a torsional angle value ~60°. In the right hand structure the value is −150° *relative to the* 60° (or 270°) and another at +150° (relative to 60°) is equivalent (or 210°). After Smith and Boyd [16] with permission.

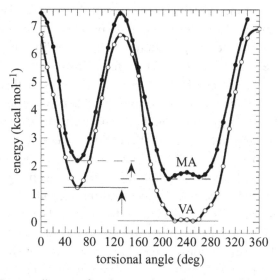

Figure 7.16 Energy diagram for the rotation of an ester side group in methyl acrylate (MA) and vinyl acetate (VA) type polymers for the situations illustrated in Figure 7.15. Stable positions at the side group orientation of 60°, and ±150° relative to 60° (or 210° and 270°) are apparent for both MA and VA. However, the energy difference (indicated by vertical arrows) between the 60° site and the ±150° sites is considerably less in the MA case than for VA. Results of Smith and Boyd [15].

example, there were no correlations between dipoles in the same chain, then the relaxation strength according to the site model exposition in Appendix AII, and in particular eq. (AII.17), would be given for the three-site picture here ($\alpha = 60°$, $\beta = 210°$, $\gamma = 270°$) as

$$N_p\, g_p\, \mu_0^2 = N\mu_0^2[2p(1)p(2)(1 - \cos\alpha)$$
$$+ 2p(2)p(3)(1 - \cos\beta) + 2p(1)p(3)(1 - \cos\gamma)], \qquad (7.1)$$

where $p(1)$, $p(2)$, $p(3)$ are simple Boltzmann factors when based on a lowest energy state of zero and, assuming the energies of 2, 3 to be the same at ΔU, are:

$$p(1) = 1/(1 + 2\exp(-\Delta U/kT),$$
$$p(2) = p(3) = \exp(-\Delta U/kT)/\exp(1 + 2\exp(-\Delta U/kT). \qquad (7.2)$$

Thus as temperature is lowered $p(2)$, $p(3)$ approach zero and the relaxation strength goes to zero. This, of course, expresses the physically obvious situation that at low temperature all of the dipoles settle into the lowest energy state and no relaxation takes place. Conversely, at a given temperature a greater value of ΔU obviously leads to diminished relaxation strength. Although the effects of dipole correlation were accounted for in the situation recounted in this section they turn out to be relatively minor and the difference in strength between methyl acrylate and vinyl acetate is largely due to site energy differences as exemplified in Figure 7.16.

The quantitative success of the picture for methyl acrylate homopolymer is expressed in Figure 7.17 and for PVAc in Figure 7.18. It is seen that good agreement with experiment is attained thus giving some confidence that the remarkable difference in relaxation strength between the methyl acrylate and vinyl acetate cases is explained molecularly in terms of conformational energies.

It has also been possible to apply molecular modeling to the dynamic aspects of subglass relaxation [17]. Two central questions arise concerning such relaxations. The first is the reason for their occurrence and the second is the reason for their very broad dynamical behavior in time or frequency. These are taken up more generally in Sec. 8.2 but some results pertinent to flexible side group motion are addressed here. These results concern the effect of interaction with its surroundings on the rotation of a methyl acrylate type side group. An energy diagram such as that in Figure 7.16 serves to establish the presence of stable sites (i.e., bond rotational angles) that are useful in rationalizing the strength behavior. In principle, it might be thought that the barriers in such diagrams could be invoked in an absolute rate theory context to determine the rates of barrier crossings and thus determine relaxation times. Carrying out such a procedure in a bulk system requires reflection. The analysis above connected with relaxation strength deals with equilibrium where the groups in question are in essentially energy minimum conformations and the surrounding molecular packing has adjusted accordingly. In the dynamic

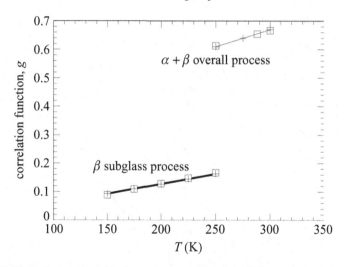

Figure 7.17 Relaxation strength as expressed by the Kirkwood–Onsager correlation function, g, (eg.(AII.10)) for methyl acrylate homopolymer. Crosses are experimental values and open squares are calculated in the subglass case from the statistical mechanical averaging of an ensemble of chains with the main–chain torsional angles fixed. The overall process case is from the same ensemble of chains but with the statistical mechanical averaging including the main chain torsional angles as well. Results of Smith and Boyd [15].

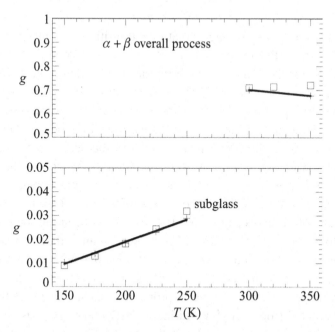

Figure 7.18 The relaxation strength in vinyl acetate homopolymer as expressed by the correlation function g. See Figure 7.17 for details. Note the factor of 10 difference in scale in the two panels representing the overall process and the subglass process. Results of Smith and Boyd [15].

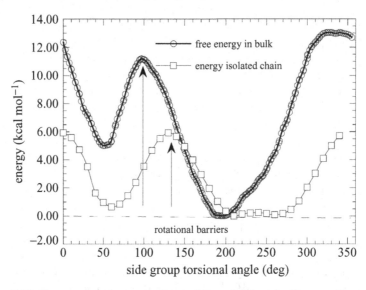

Figure 7.19 Free energy curve constructed from MD umbrella sampling of the rotation of a single methyl acrylate side group attached to a poly(ethylene) chain embedded in a matrix of poly(ethylene) chains at 200 K. Shown for comparison is the energy curve in Figure 7.16 for a methyl acrylate group on an isolated poly(ethylene) chain. Results of Smith and Boyd [17].

situation, a side group undergoing a rotational transition not only experiences the intramolecular energetics embodied in a diagram like Figure 7.16 but no doubt also encounters repulsive interactions with the surrounding molecules. Although an energy diagram may exist for a given path it presumably is strongly influenced by the surrounding packing. The question is then whether variations in packing environments can lead to variations in effective barriers in bond rotational jumps that are sufficient to lead to the broad relaxation processes observed experimentally. Straightforward application of MD simulation can be applied but at the low temperatures characteristic of the glass very long trajectories would be needed to observe "spontaneous events;" i.e., a rotation conformational jump of a side group. Since the nature of the expected jump is known, i.e., a side group rotation, MD can be adapted to ensure such transitions. This involves umbrella sampling in which the desired motion is forced by applying an artificial potential that biases the state of the group in question. In practice, the biasing potential chosen is effective over a fairly narrow range of torsional angle. A probability density curve for the torsional angle is constructed from sampling during an MD trajectory. It is then matched with other biasing windows by overlap in the windows. Figure 7.19 shows the free energy derived from a set of overlapping probability functions determined from MD sampling for a single methyl acrylate side group attached to a poly(ethylene) chain embedded in a matrix of poly(ethylene) chains [17].

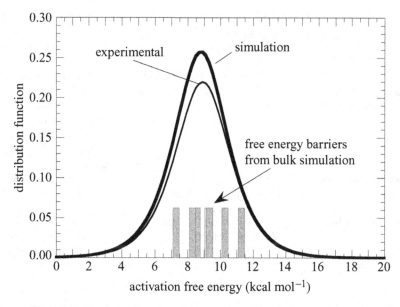

Figure 7.20 Distribution of activation energies. The free energy barriers from simulations of seven distinct glass structures are marked on the abscissa as vertical bars. The curve marked simulation is a Cole–Cole distribution function with the same average activation energy and standard deviation as the seven simulations. The experimental curve is a Cole–Cole curve calculated from measured α, τ_0 parameters for dilute methyl acrylate/poly(ethylene) copolymers [11]. Results of Smith and Boyd [17].

It is apparent in Figure 7.19, but not surprising, that the effect of the surrounding matrix is considerable. Both the position of the barrier and its height are affected. The question of whether variation in barrier heights in various environments leads to the observed very broad relaxations is addressed in the results in Figure 7.20.

MD umbrella sampling free energy curves were generated for a total of seven distinct packing environments. The latter were prepared by starting from cooling of liquid systems at high temperature that were distinct by virtue of having had differing equilibration times. The side group rotational barriers for the seven systems are marked by the vertical bars along the abscissa. It is seen that there is considerable variation in the barriers. Comparison with experiment can be made as follows. The distribution of relaxation times for the Cole–Cole function, eq. (2.49), is reprised for unit strength as

$$\overline{F}(\ln \tau) = \frac{1}{2\pi} \left[\frac{\sin \pi \alpha}{\cosh(\alpha \ln(\tau/\tau_0)) + \cos \pi \alpha} \right]. \tag{7.3}$$

The temperature dependence of relaxation times is assumed to be Arrhenius in nature or,

$$\ln(\tau(T)) = \Delta H(\tau)/RT + B \tag{7.4}$$

Table 7.1. *Summary of mechanical and dielectric subglass process characteristics*

Descriptor	Mechanical (M)	Dielectric (D)
Strength (relaxed vs. unrelaxed modulus or dielectric constant)	highly variable, but much less than glass transition region on log $E(t)$ basis	highly variable, usually less than glass transition region, depends on differences in energy minima of conformational states of participating groups
Temperature dependence of location	Arrhenius ΔH^* modest 10–20 kcal mol^{-1} $\Delta H^*(M) \sim \Delta H^*(D)$	Arrhenius ΔH^* modest 10–20 kcal mol^{-1}
Shape vs. log t or log f	extremely broad (but no $E_r \ll E_u$ effect)	extremely broad, much broader than primary glass transition region

and the constant B is assumed to be the same for all relaxation times. This allows the introduction of $\Delta H(\tau)$ as the independent variable in eq. (7.3), or,

$$\overline{F}(\Delta H) = \frac{1}{2\pi RT} \left[\frac{\sin \pi \alpha}{\cosh(\alpha(\Delta H - \Delta H_0)/RT) + \cos \pi \alpha} \right]. \qquad (7.5)$$

This equation is used to construct a complete distribution of relaxation times that has the same standard deviation and average activation energy as the vertical bars in Figure 7.20. This function is compared to an experimental one, the latter being based on Cole–Cole function α width parameters and τ relaxation times from measurements on methyl acrylate/poly(ethylene) copolymers that are chemically dilute in methyl acrylate units. The two distributions are very similar, a situation that lends credence to the proposition that relaxation breadth is related to the premise that activated conformational jumps experience barriers that are highly variable due to the vagaries of packing environments (Table 7.1).

7.4 NMR spectroscopy studies of flexible side group motion

The exchange NMR technique centerband-only detection of exchange (CODEX), which is a variation of the 2D exchange methods introduced in Chapter 3, enables detailed and quantitative studies of slow (milliseconds to 10 seconds) motions in complex polymers without isotopic labeling [18,19]. This method is particularly powerful when applied to polymers with flexible side groups as it allows the determination of the fraction of groups that rotate in a given time frame as well as the amplitude of the motion of the flexible groups. CODEX has been applied to

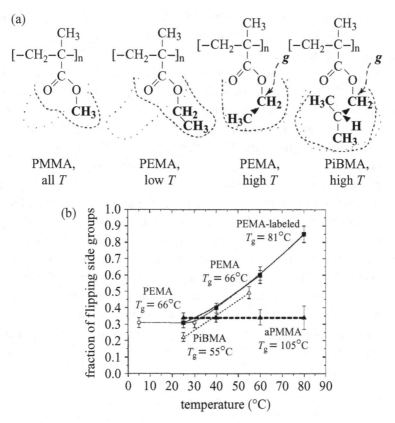

Figure 7.21 (a) Representations of side group conformations for PMMA, PEMA and PiBMA. The dotted lines indicate the space required for the side group after a 180° flip. A hypothetical high temperature *gauche* conformation, which makes the side group more symmetric and favors side group flips, is shown for PEMA, and PiBMA. (b) Temperature dependence of the fraction of flipping side groups in PMMA, PEMA, and PiBMA (open triangles) from CODEX measurements. Taken from Bonagamba *et al.* [20] with permission.

the study of side group motion in a series of glassy poly(alkyl methacrylates), including poly(methyl methacrylate) (PMMA), poly(ethyl methacrylate) (PEMA) and poly(isobutyl methacrylate) (PiBMA), as illustrated in Figure 7.21(a) [20]. The β-relaxation in these polymers is believed to involve large-amplitude motion of the flexible ester side groups. CODEX reveals that in the glassy state the side groups of these polymers undergo 180° rotations, or flips, as illustrated in Figure 7.21(a). However, as summarized in Figure 7.21(b), only a fraction of the ester side groups are active, revealing that the fraction of mobile segments is an important fundamental aspect of relaxations in glassy polymers involving flexible side group motion in addition to the rate and amplitude of the motion.

The CODEX measurements on glassy poly(alkyl methacrylates) indicate that the temperature dependence of the fraction of active (flipping) side groups depends upon the chemical structure of the side group. Specifically, Figure 7.21(b) shows that, while the fraction of active side groups in PMMA is independent of temperature below T_g, it increases dramatically in PEMA and PiBMA with increasing temperature to the point where the active fraction of these larger ester side groups becomes greater than that for the smaller group in PMMA. As shown in Figure 7.21(a), the side group in PMMA has an asymmetric shape that is temperature independent due to a lack of internal conformational flexibility. The shape asymmetry can lead to packing asymmetry and hence inaccessibility (high energy) of the 180° rotated configuration for any given side group due to steric crowding. This crowding effect could explain why only a fraction of the ester side groups are active in the poly(alkyl methacrylate) glasses. While shape asymmetry exists for the low energy conformation of the ester side group in PEMA (see Figure 7.21(a)) and PiBMA, these larger ester groups are conformationally flexible. Higher energy but more symmetric side group conformations become more accessible with increasing temperature, perhaps facilitating flips as illustrated in Figure 7.21(b) and thereby resulting in an increase in the fraction of active side groups with increasing temperature [20].

References

[1] K. H. Illers and H. Breuer, *J. Colloid Sci.*, **18**, 1 (1963).
[2] R. H. Boyd and F. Liu in *Dielectric Spectroscopy of Polymeric Materials*, edited by J. P. Runt and J. J. Fitzgerald (Washington, DC: American Chemical Society, 1997) Chapter 4.
[3] N. G. McCrum, B. E. Read, and G. Williams, *Anelastic and Dielectric Effects in Polymeric Solids* (New York: Wiley, 1967; Dover, 1991). Data of H. Takayanagi.
[4] S. U. Boyd and R. H. Boyd, *Macromolecules*, **34**, 7219 (2001).
[5] S. P. Bravard and R. H. Boyd, *Macromolecules*, **36**, 741 (2003).
[6] T. A. Ezquerra, F. J. Balta-Calleja, and H. Zachmann, *Acta Polym.*, **44**, 18 (1993).
[7] R. M. Fuoss, *J. Am. Chem. Soc.* **63**, 369 (1941); **63**, 378 (1941).
[8] J. Larsson, Master of Science Thesis, Royal Institute of Technology, Stockholm (1999).
[9] O. Johansson and T. Falk, Master of Science Thesis, Royal Institute of Technology, Stockholm (1993).
[10] Y. Ishida, M. Matsuo, and K. Yamafuji, *Kolloid Z.*, **180**, 108 (1962).
[11] D. E. Buerger and R. H. Boyd, *Macromolecules*, **22**, 2694 (1989).
[12] D. E. Buerger and R. H. Boyd, *Macromolecules*, **22**, 2699 (1989).
[13] R. W. Devereaux, B.Sc. Thesis, University of Utah (1988).
[14] R. H. Boyd and F. Liu, previously unpublished results.
[15] G. D. Smith and R. H. Boyd, *Macromolecules*, **24**, 2731 (1991).
[16] G. D. Smith and R. H. Boyd, *Macromolecules*, **24**, 2725 (1991).

[17] G. D. Smith and R. H. Boyd, *Macromolecules*, **25**, 1326 (1992).

[18] E. R. deAzevedo, W.-G. Hu, T. J. Bonagamba, and K. Schmidt-Rohr, *J. Am. Chem. Soc.*, **121**, 8411 (1999).

[19] E. R. deAzevedo, W.-G. Hu, T. J. Bonagamba, and K. Schmidt-Rohr, *J. Chem. Phys.*, **112**, 8988 (2000).

[20] T. J. Bonagamba, F. Becker-Guedes, E. R. deAzevedo, and K. Schmidt-Rohr, *J. Polym. Sci.: Part B: Polym. Phys.*, **39**, 2444 (2001).

8

The transition from melt to glass and its molecular basis

8.1 Experimental description

The preceding chapters of Part II have dealt with the primary relaxation that takes place in the melt and secondary subglass relaxations at lower temperature. However, it is a matter of experimental observation that the time or frequency locations of these relaxations tend to merge as temperature increases or conversely to diverge as temperature is lowered. This is implied in loss maps such as Figure 7.13 where extrapolation to higher temperature, as illustrated in Figure 8.1, indicates the secondary and primary relaxations merge. In fact the extrapolations based on the centers of the processes seem to converge in the region of $\sim 10^{11}-10^{12}$ Hz or on the time scale of picoseconds. This would seem to be inevitable since the time scale of slower intramolecular vibratory and torsional oscillations is of this order and motions related to more elaborate conformational transitions could not be faster than these limits. Since the relaxations are broad, the merging region is expected to be noticeable experimentally in loss vs. frequency curves at temperatures well below this implied high frequency limit. In fact this is the case.

8.1.1 Examples of the overlap and coalescence behavior from dielectric spectroscopy

It is useful to recognize several stages in the merging. The first is an overlap region where maxima in the loss processes for both relaxations are clearly visible. Because the secondary process is often, but not always, broader and thus has a smaller maximum in the dielectric loss there is a coalescence region where there is a shoulder signified by an inflection point in the loss curve. Finally, there is the completely merged region.

The data in Figure 8.2 show the overlap region in amorphous PET. The curves in the figure are fits based on two additive HN functions, one for the α-process

142

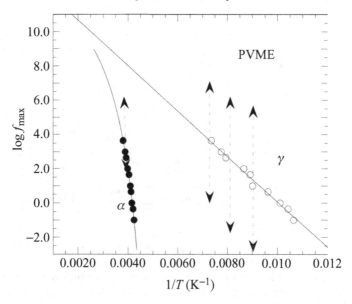

Figure 8.1 Loss map of Figure 7.13 with extrapolated higher temperature/ frequency behavior (solid curves). The vertical lines indicate the frequency of breadth of the loss at 20% of the maximum loss.

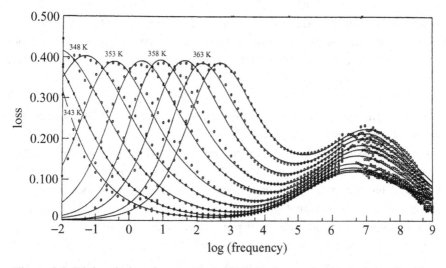

Figure 8.2 Dielectric loss in amorphous PET in the overlap region showing the approach to coalescence of the α- and β-relaxations as temperature increases. The curves are HN function fits of the data. Taken from Hoffmann *et al.* [1] with permission.

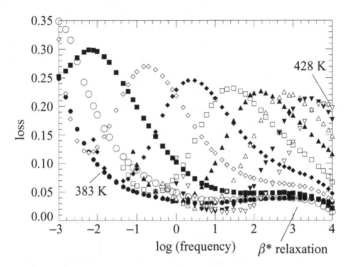

Figure 8.3 Dielectric loss of PEN in the α-process region including the merging with the β^*-relaxation as temperature increases. The curves are at 5 K intervals. Data of Bravard and Boyd [2].

and one for the β-process. The actual coalescence and complete merging cannot be followed since amorphous PET crystallizes before this region can be approached. However, in comparison with PET, PEN crystallizes at a higher temperature relative to the primary (α) and secondary processes (β^*, β, Figure 7.7). As seen in Figure 8.3, this allows the entire sequence of almost separate processes (383 K), overlapping (393 K), coalescence (398, 403 K), and complete merging (408 K) of the α- and β^*-processes to be observed.

It is appropriate to note that the details of the merging are dependent on the relative strengths of the processes. It was commented previously that the secondary relaxation in PMA is exceptionally strong. The dielectric loss spectrum for the secondary β-relaxation in PMA at low temperatures, where it is well resolved from the α-process, was shown in Figure 7.14. The spectrum at much higher temperatures and frequencies, where the β-relaxation merges with the α-process, is displayed in Figure 8.4. It can be seen that the β-relaxation narrows and that the α-relaxation actually becomes a shoulder on the β-process in the final merger.

It is of interest to know the extent to which the overlapping processes may be regarded as more or less independent and hence as additive in their contributions to the overall process. Certainly, from a curve fitting standpoint, it is likely that parameters can be chosen for additive phenomenological functions, such as two HN functions, that represent the measured isothermal loss vs. frequency data well. The real question is whether the fitting parameters represent a reasonable extension vs. temperature of the parameters for the separated processes at lower temperature.

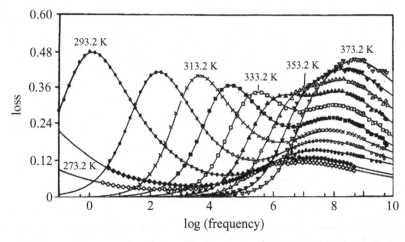

Figure 8.4 Dielectric loss in PMA over a broad frequency range that shows the coalescence of the α- and β-relaxations as temperature increases. The curves smooth the data. From Schoenhals [3] with permission.

In the case of the PEN data in Figure 8.3, the loss data in the vicinity of the α-process peaks were fit to an HN function but with the high frequency tail data not incorporated in the fitting process [2]. Figure 8.5 shows the result for the 398 K data in the coalescence region. Also shown are the result of subtracting the HN curve fit from the data points, leaving an implied contribution of the β^*-process and the location of the β^*-process loss peak expected from Arrhenius extrapolation of the β^*-process at lower temperatures. It can be seen that the location agrees well with the loss peak implied by the curve fitting in Figure 8.5. In general, in the overlap region the measured properties are consistent with additive behavior and in this case in the coalescence region as well.

An example in which numerical regression has been applied jointly to α- and β-processes in the overlap and coalescence regions is shown in Figure 8.6 [4]. The polymer involved is a linear aliphatic random copolyester where the diol unit is ethylene glycol and the acid moieties are adipic and succinic acids in $1:1$ mole ratio (acronym PE-S/A). The polymer thus has only main chain relaxation ability, possesses significant polarity, and, especially, avoids the complication of crystallization. At low temperatures the subglass β-process shows structure, in a vein similar to PET and PEN, but that can be represented by overlapping β_1- and β_2-processes. However, in the temperature region of overlap and coalescence the two β-relaxation components have merged into a single process. Figure 8.7 shows the results of numerical regression (or "unbiased modeling" [5]) based on two additive functions, one for the α-process (HN) and one for the β-process (Cole–Cole) for the data in Figure 8.6 at the four temperatures where the frequency window captures

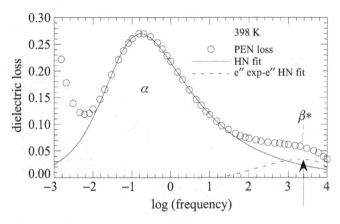

Figure 8.5 Fit to the dielectric loss in the PEN α, β coalescence region using the HN function. The dashed curve is the difference between the data and the HN function and thus is the implied β^*-process contribution. The vertical arrow indicates the location of the β^*-process log f_{max} from extrapolation of the experimental values at lower temperature. Results of Bravard and Boyd [2].

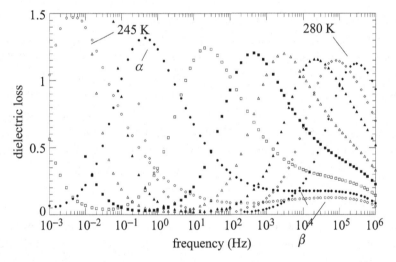

Figure 8.6 Dielectric loss peaks in an aliphatic copolyester (see text) in the α- and β-process overlap and coalescence regions (from 245 K to 280 K at 5 K intervals). Data of Sen [4].

the merging process. In any numerical regression procedure it is to be hoped that there is a degree of uniqueness in the determined parameter values, i.e., that they are reasonably robust with respect to differing starting estimates. That appears to be the case in the present example.

Returning to the question of the degree to which the determined parameters reflect additive process behavior or more complicated behavior, the temperature

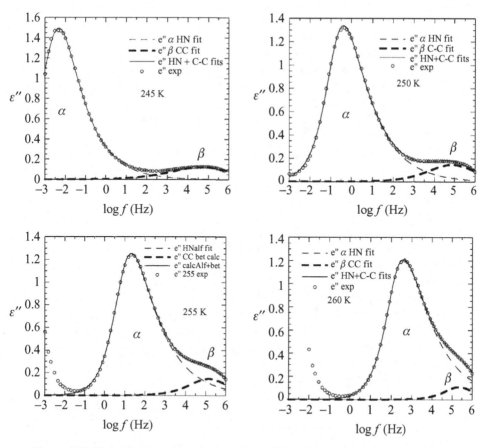

Figure 8.7 Numerical regression fits of the dielectric loss peak in an aliphatic copolyester (see text) at four temperatures spanning the α- and β-processes overlap and coalescence regions. Two functions were used at each temperature, an HN function for the α-process and a Cole–Cole function for the β-process. The overall regression fit is indicated by the solid line and the individual contributions of the α-process HN and β-process Cole–Cole functions are also shown. Results of Sen [4].

dependence of the central relaxation times of the α- and β-processes determined by the regressions is shown in the upper panel of Figure 8.8 for the PE-S/A copolymer. The relaxation times for the lower temperature range β_1- and β_2-processes are also shown along with the complete range of measured α-process $\log f_{\mathrm{max}}$ values. The temperature behavior of the β-relaxation times in the overlap and coalescence regions is seen to be consistent with an extrapolation of the lower temperature values. However, the situation is more complicated with respect to the width parameters as displayed in Figure 8.8(b). In the overlap region at 245 K, the width parameter for the Cole–Cole representation of the β-process is consistent

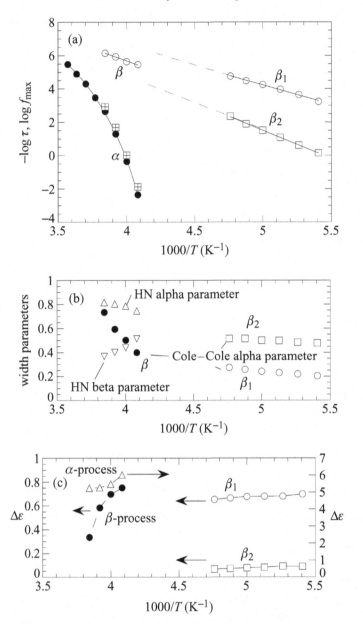

Figure 8.8 Regression parameters for the fits to the overlap and coalescence regions in the PE-S/A copolyester fits in Figure 8.7. (a) The relaxation times for the α-process (squares with crosses) and the β-processes (open circles for β, β₁ and squares for β₂): also shown (as filled circles) are the log f_{max} values for all of the $ε''$ peaks in Figure 8.6. (b) The width parameter results: the open circles and squares are the Cole–Cole width parameters for the β₁, β₂ components of the β-processes at low temperature; the filled circles are the Cole–Cole width parameters for the β-process in the coalescence region; the triangles are the HN α-process width parameters. (c) The dielectric strength increments: the arrows indicate which ordinate scale is used. Results of Sen [4].

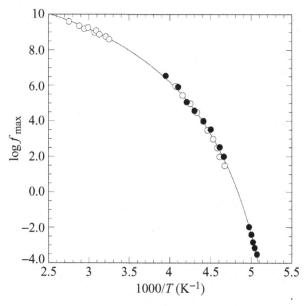

Figure 8.9 Loss peak map for PPO. The curve is a VF fit (eq. (6.7)) where $A = 12.23$, $B = 514.2$, $T_\infty = 164.9$. Filled points are data of Williams [8]; the open points are data of Yano *et al.* [9].

with the lower temperature β_1- and β_2-processes, but as temperature increases the β-process sharpens dramatically as indicated by the increasing width parameter. Simultaneously, the α-process β skewing parameter *decreases* (increased skewing). In addition, as seen in Figure 8.8(c), where the relaxation strengths are shown, the α-process strength remains relatively constant and the β-process strength decreases as coalescence proceeds with increasing tempreature. To summarize, coalescence results in the higher frequency β-process broadening the combined processes toward high frequency. Numerical regression accommodates this as increased skewing in the lower frequency but a much stronger α-process (even though HN skewing typically decreases somewhat with increasing temperature in well separated α-processes). The β-process is reduced to a weaker, narrower process.

High temperature, high frequency behavior

The behavior of the α-relaxation process (or merged processes if there is a significant secondary process) as temperature increases further is also of interest. Figure 8.9 shows a loss peak map for poly(propylene oxide) (PPO) that encompasses a very wide frequency range. In Figure 8.10 a series of Argand plots is displayed. The lowest temperature one represents the middle frequency region of the loss map

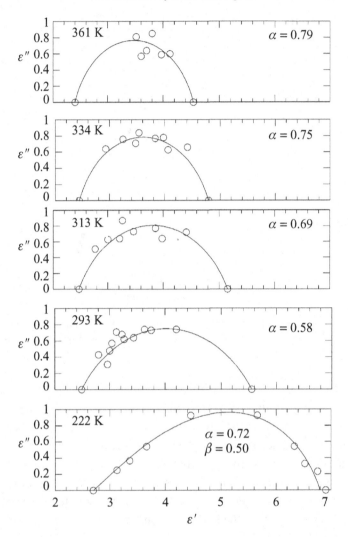

Figure 8.10 Argand diagrams at a number of temperatures for the primary relaxation in PPO. The lowest temperature corresponds to the intermediate temperature region in Figure 8.9 and the fitting curve is the HN function with the parameters indicated. The plots at the other temperatures are from the high frequency region of Figure 8.9 and the fitting curves are Cole–Cole functions with the listed width parameters. Data of Yano *et al.* [9].

in Figure 8.9. This plot shows the HN function behavior typical of a primary α-relaxation process. The rest of the plots are from the high frequency region (up to ~7 GHz). It is seen the process loses its HN asymmetry at higher temperature and Cole–Cole functions are used to represent data. As temperature rises, although the relaxation strength diminishes, the process becomes narrower with a corresponding increase in the Cole–Cole function α width parameters. In fact, at the highest temperature, SRT behavior is approached. The relaxation strength decrease is a simple consequence of the "$1/T$" effect when the correlation factor (eq. (AII.10)) is otherwise rather insensitive to temperature. The narrowing and approach to SRT behavior is noted in other polymers (e.g., melts of Nylon 610 [6], poly(ethylene oxide) and poly(oxymethylene) [7]), at very high frequency as well and apparently is a general phenomenon.

8.1.2 Neutron scattering and NMR exchange measurements

Neutron spin echo and the β-relaxation

As discussed in Sec. 8.1.1, temperatures at which both the primary α- and secondary (main chain) β-relaxation processes in amorphous polymers are observable isothermally through dielectric spectroscopy are usually limited to a relatively narrow range near T_g. At temperatures sufficiently low to observe well separated processes the β-relaxation time is on the order of microseconds or longer. Because the accessible time window of dynamic neutron scattering is currently limited to times less than 100 ns it would seem that dynamic neutron scattering techniques are not applicable to the study of the β-relaxation in amorphous polymers. However, the possibility that the β-relaxation may be observable through dynamic neutron scattering at higher temperatures, and hence shorter relaxation times, has motivated efforts to utilize dynamic neutron scattering as a probe of the β-relaxation and the bifurcation of the α- and β-processes [10,11].

Figure 8.11 shows the coherent dynamic structure factor $S_{coh}(q, t)$ for the same PBD melt discussed in Sec. 6.5 obtained from NSE measurements with $q = 2.71$ Å^{-1}, corresponding to the second peak in the static structure factor (Figure 6.26(b)) and hence reflecting entirely intramolecular correlations. For $q = 1.48$ Å^{-1} it was found that $S_{coh}(q, t)$ obtained at different temperatures could be made to superimpose by scaling time by the temperature dependent viscosity, as shown in Figure 6.26(a), indicating that structural relaxation and melt viscosity follow the same VF temperature dependence. As seen in Figure 8.11, scaling time for $q = 2.71$ Å^{-1} by the temperature dependent viscosity does not yield a master relaxation curve, indicating that polymer motions probed at $q = 2.71$ Å^{-1} exhibit a different dependence on temperature than those observed at $q = 1.48$ Å^{-1} (structural relaxation) or exhibited by the macroscopic viscosity.

Fitting $S_{coh}(q, t)$ for $q = 2.71$ Å^{-1} at each measurement temperature with a KWW stretched exponential using a stretching exponent given by dielectric relaxation ($\beta = 0.41$) in a manner analogous to that employed for $q = 1.48$ Å^{-1}

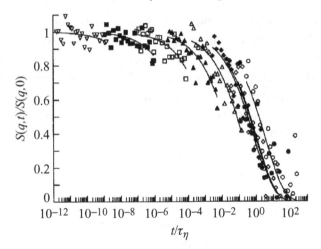

Figure 8.11 Scaling representation of NSE data for PBD for $q = 2.71$ Å^{-1} for temperatures ranging from 170 K (open triangles) to 300 K (open circles). Solid lines correspond to KWW functions (eq. (2.37)) with $\beta = 0.41$. Taken from Arbe *et al.* [11] with permission.

(Figure 6.26) yields the temperature dependent relaxation times shown in Figure 6.26(c). The temperature dependence of the relaxation times obtained from NSE measurements at $q = 2.71$ Å^{-1} can be described reasonably well by the Arrhenius temperature dependence of the dielectric β-relaxation, as shown in Figure 6.26(c), supporting the supposition that while $S_{\text{coh}}(q, t)$ at $q = 1.48$ Å^{-1} probes the α-relaxation, i.e., the same motions that ultimately determine macroscopic relaxation behavior, $S_{\text{coh}}(q, t)$ at $q = 2.71$ Å^{-1} is sensitive to the same localized motions that lead to the dielectric β-process observed at lower temperatures. However, extrapolating the relaxation times for the neutron "β-process" (Figure 6.26(c)) to the temperature regime of the resolvable dielectric β-process in PBD yields relaxation times two orders of magnitude smaller than observed from dielectric relaxation measurements.

In order to resolve this apparent contradiction, MD simulations of the PBD melt introduced in Sec. 6.4 have been performed over a wide temperature range [12]. $S_{\text{coh}}(q, t)$ was obtained for the PBD melt at each simulation temperature through Fourier transformation of the van Hove correlation functions and is shown for $q = 2.72$ Å^{-1} in Figure 8.12(a). The initial decay ($t < 2$ ps), due primarily to motion associated with torsional librations, is quite significant at this large q value and accounts for 90% of the decay of $S_{\text{coh}}(q, t)$ at high temperatures. At low temperatures, $S_{\text{coh}}(q, t)$ exhibits a plateau beginning at around 2 ps that extends to longer time with decreasing temperature. This plateau signifies the development of another dynamic regime between the microscopic dynamics and the long-time structural or

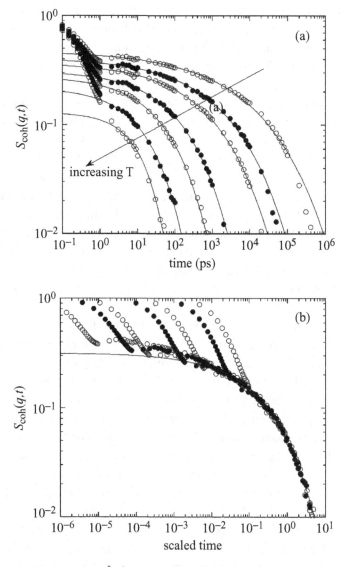

Figure 8.12 $S_{coh}(q = 2.72 \text{ Å}^{-1}, t)$ for a PBD melt from atomistic MD simulations at several temperatures ranging from 198 K to 293 K. (a) The solid lines are KWW fits to the data for $t \geq 2$ ps. (b) $S_{coh}(q, t)$ vs. scaled time. The solid line is a KWW fit to the α-relaxation regime. Data from Smith *et al.* [12].

α-relaxation that leads to the final decay of $S_{coh}(q, t)$. The plateau is connected with the caging process of monomers and, within the mode coupling theory (MCT) of the glass transition, is described as a separate dynamic regime, the MCT β-relaxation [13]. In order to investigate the applicability of time–temperature superposition for the α-relaxation, the time axis for each temperature was scaled by the time $t^*(T)$ to reach a fixed value, specifically $S_{coh}(q, t^*(T)) = 0.05$ for $q = 2.72$ Å^{-1}, as shown in Figure 8.12(b). The α-relaxation dominated tails of $S_{coh}(q, t)$ superimpose over a wide range of temperature indicating that time–temperature superposition for the α-relaxation is obeyed. $S_{coh}(q, t)$ for $q = 2.72$ Å^{-1} and $q = 1.44$ Å^{-1} (not shown here) in the α-relaxation regime were fit with KWW stretched exponential with stretching exponents of $\beta = 0.382$ and 0.617, respectively. The representation of the master curve for the α-relaxation for $q = 2.72$ Å^{-1} is shown in Figure 8.12(b). An excellent representation of $S_{coh}(q, t)$ is obtained in the α-relaxation regime. At short times, significant deviation of $S_{coh}(q, t)$ from the KWW fit to the α-relaxation can be seen for all temperatures. The initial short-time decay ($t < 2$ ps) (see Figure 8.12(a)) due to librational motion, i.e., dihedral motion of the polymer within conformational energy wells, does not obey the time–temperature superposition observed for the α-relaxation and cannot be described by the KWW representation of the α-relaxation. Deviation between $S_{coh}(q, t)$ and the KWW fit to the α-relaxation can be seen for times short compared to the α-relaxation time and longer than the initial short-time decay. This deviation can be associated with the MCT α-relaxation [13].

The relaxation times of the α-relaxation, obtained from the KWW fits to the simulation data, are shown as a function of inverse temperature in Figure 8.13 for $q = 1.44$ Å^{-1} and $q = 2.72$ Å^{-1}. The α-relaxation times for both q values can be well represented by a VF temperature dependence. The temperature dependence of the α-relaxation time extracted from $S_{coh}(q, t)$ does not exhibit significant dependence on q, supporting the conclusion that the underlying physical processes (molecular motions) responsible for the relaxation of $S_{coh}(q, t)$ are the same on both length scales ($q = 1.44$ Å^{-1} and $q = 2.72$ Å^{-1}) examined. Figure 8.13 also shows a comparison of the reported relaxation times at $q = 1.48$ Å^{-1} and $q = 2.71$ Å^{-1} obtained from NSE measurements with values obtained from simulation at $q = 1.44$ Å^{-1} and $q = 2.72$ Å^{-1}. While good correspondence of the time scales for the α-relaxation between simulations and experiment can be seen, systematic deviation between simulation and experiment in the temperature dependence of the relaxation times for the larger q, particularly at lower temperatures, can be observed.

In order to account for this discrepancy $S_{coh}(q, t)$ at $q = 2.72$ Å^{-1} from simulation was refit with a KWW expression with $\beta = 0.41$ with fitting restricted to the experimental time window (2 ps–2 ns), emulating the procedure utilized to extract relaxation times from NSE measurements. The resulting correlation times are shown in Figure 8.13. It can be seen that increasingly large deviations from the true α-relaxation time emerge with decreasing temperature. Figure 8.13 reveals that the relaxation times obtained by applying the experimental procedure to the simulation $S_{coh}(q, t)$ can be fit with an Arrhenius temperature dependence that also

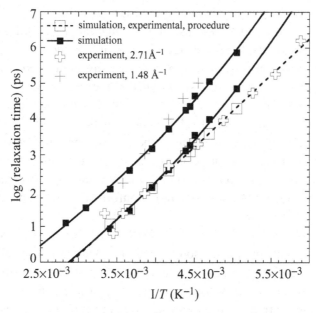

Figure 8.13 Comparison of α-relaxation times obtained from simulation and analysis of NSE measurements on PBD. The solid lines are fits to the VF equation using $C_2 = 1125$ K, $T_0 = 72$ K. The dashed line at $q = 2.72$ Å^{-1} is an Arrhenius fit to the simulation relaxation times obtained using the experimental procedure. Simulation data from Smith *et al.* [12]; experimental data from Arbe *et al.* [11].

provides a very good description of the experimental data. Hence, it appears that the claimed difference in the temperature dependence of the α-relaxation times obtained for the two q values from NSE measurements [11] is an artifact of the procedure used in analyzing the experimental NSE data that becomes increasingly important with decreasing temperature. Simulations reveal that in order to obtain accurate relaxation times one must utilize data that adequately sample the α-relaxation regime. Relaxation times extracted from data over a too-narrow time window show systematic deviations from the true relaxation times. The NSE measurements for $q = 2.72$ Å^{-1} sample the α-relaxation ($S_{\mathrm{coh}}(q, t) < 0.2$) only for the higher temperatures measured. For the lower temperatures, where significant deviation from simulations that access much longer times than the NSE measurements is seen, experiments have not even begun to measure motions associated with the α-relaxation process.

Multidimensional NMR exchange measurements and dynamic heterogeneity near T_g

Multidimensional exchange NMR studies of polymer dynamics near the glass transition temperature have been instrumental in demonstrating that the non-exponential time behavior observed for polymer relaxations is related to a superposition of relaxation processes, each characterized by an individual rate [14,15]. NMR exchange measurements reveal that the α-relaxation process in polymers near the glass transition temperature is dynamically heterogeneous, being composed of a distribution of relaxing segments ranging from slow to fast. For example, accurate representation of the 2D solid-state exchange ^{2}H spectrum of PS shown in Figure 3.8 requires a distribution of relaxation times for local conformational motion with a width of several decades. Furthermore, modeling of the 2D exchange spectra for PS obtained near the glass transition temperature has revealed that the width of the reorientation time distribution for C–^{2}H vectors increases dramatically with decreasing temperature, consistent with the MD simulations of conformational dynamics in polymers near T_g discussed in Sec. 8.2.

Two-dimensional solid-state exchange NMR studies have also led to the conclusion that near T_g local polymer motion occurs through a broad distribution of reorientational angles indicating ill-defined reorientation motions with strong diffusive components. It was concluded that the importance of large-angle conformational changes decreases with decreasing temperature and that *trans–gauche* transitions do not play a dominant role in C–^{2}H reorientation close to the glass transition temperature of ps [16]. Similar conclusions were drawn for poly(isoprene) [14] and atactic poly(propylene) [17]. These results are in conflict with MD simulation studies of a number of polymer melts that indicate that conformational transitions remain the dominant mode of local motion in polymers through the glass transition and, in fact, in the glass itself, as discussed in Sec. 8.2.

While the 2D solid-state exchange spectra reveal an increasingly broad distribution of relaxation times for motions associated with the α-relaxation as the temperature is reduced toward T_g, these studies say nothing about the time scale of fluctuations within the distribution of reorientation rates. In other words, these studies say nothing about the time scale over which slowly reorienting segments remain slow and fast reorienting segments remain fast. Reduced 4D solid-state exchange measurements [14,18] have provided important insight into this question, revealing that fluctuations within the rate distribution occur on the same time scale as the α-relaxation processes itself, consistent with the results from MD simulations discussed in Sec. 8.2. Finally, efforts have been made to address the question of whether the dynamic heterogeneity observed in motions leading to the α-relaxation is spatially correlated, i.e., whether there are spatial domains of slow and fast relaxing segments. Perturbative methods, such as confining liquids to pores or introducing large probe molecules, appear to indicate spatial correlations on the length scale of a few nanometers within 20 K of T_g [19]. A non-perturbative variation of the 4D solid-state exchange method involving proton spin diffusion [15] has been employed to study PVAc slightly above the glass transition. Complex modeling of the spectra was found to indicate spatial correlation of segmental relaxation rates with a length scale of about 3 nm. While consistent with other probes of simple glass forming liquids and polymers, these results are apparently inconsistent with MD simulations as discussed in Sec. 8.2.

8.2 Molecular basis

The above experimental descriptions have largely been couched in terms of the effects of increasing temperature, i.e., overlapping, coalescing, and merging. From a conceptual viewpoint it is perhaps more useful to think in terms of vitrification, the changes taking place on cooling. This would start with the melt where the time scale of molecular events is rapid, then the bifurcation region where the presence of faster and slower processes becomes experimentally noticeable, and finally the glass where only the fast processes remain with experimental signatures as relaxations. This view is especially suited to molecular interpretation via simulations. MD simulations, in general, are useful through capturing autocorrelation functions (ACFs) for the decay of various properties that hopefully are measurable experimentally and monitoring at the molecular level local events such as conformational transitions (i.e., bond rotations) that are responsible for relaxation behavior. One of the most striking features of the vitrification process is that polymer dynamics becomes spatially heterogeneous. That is, the bonds in chemically identical units in different parts of the chain and system tend to undergo conformational transitions at greatly different rates. These individual differences can persist over time scales that are long enough to be readily observed in simulation and in appropriate experiments. The subject of heterogeneous dynamics is taken up here. Another important issue is to what extent can the bifurcation–coalescence phenomena be detected and analyzed in simulation. This is also taken up.

8.2.1 Conformational phenomena and the onset of dynamic heterogeneity

The first aspect of dynamic heterogeneity considered is its description and how it is connected to conformational events as determined via simulation. It is useful to review the investigation through MD simulation of a simple but physically real polymeric system, namely linear poly(ethylene), over the temperature range where the transition from melt to glass takes place [20–28].

Amorphous poly(ethylene) as a model system

Poly(ethylene) is an easily crystallizable polymer and crystallization can be effected and simulated via MD under the conditions of crystallization from solution [29,30,31] and crystallization under mechanical deformation [32,33]. However, under quiescent bulk conditions, the necessity of nucleation and growth render crystallization unlikely and MD simulation of amorphous poly(ethylene) can be

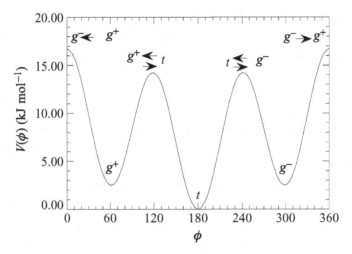

Figure 8.14 Torsional potential invoked in PE simulation. The *trans* (*t*) and *gauche* (g^+, g^-) states and the barriers between them are labeled. The $t \rightarrow g^+$, g^- and $g^- \rightarrow g^+$ barriers are both 14.2 kJ mol^{-1}. From the potential parameters of Boyd *et al.* [21].

carried out. The non-availability below the crystalline melting point of a physical amorphous poly(ethylene) does somewhat complicate comparison with experiment but as seen below useful comparisons can be made using experimental data for semi-crystalline specimens.

The system is simplified by representing the $-CH_2-$ units as a single force center. This united atom (UA) approach greatly reduces the computational effort and allows much longer time trajectories (Sec. 5.4.1). It is refined in much of the work described here by placing the force center not at the C center but displaced toward the H atoms. The local anisotropy introduced thus results in an anisotropic united atom model (AUA) [34,35]. The torsional or bond rotation potential invoked [20,21] is shown in Figure 8.14.

First, it is noted that a common signature of glass formation, a break in the slope of the volume versus temperature dependence, is observed in simulation (as is seen in Figure 8.15). In general it is found that NPT MD simulations give volumetric glass transitions that, although somewhat displaced to higher temperature, are in reasonable accord with experimental ones [36,37,38].

It is important to establish that the relaxational behavior from the MD simulations corresponds with experiment. This is accomplished via ACFs. The TACF in normalized form, eq. (5.3), is recapitulated here but with angle values in cosine form as

$$\text{TACF}(t) = \frac{(\langle \cos \phi(0) \cos \phi(t) \rangle - \langle \cos \phi(0) \rangle^2)}{(\langle \cos^2 \phi(0) \rangle - \langle \cos \phi(0) \rangle^2)}, \quad (8.1)$$

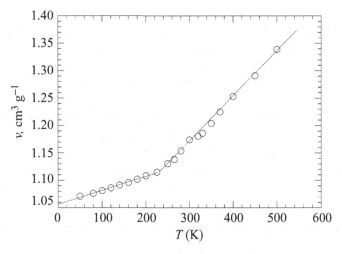

Figure 8.15 Volume dependence of PE on temperature, showing a T_g break, as determined by NPT MD simulation. Data from Boyd *et al.* [21].

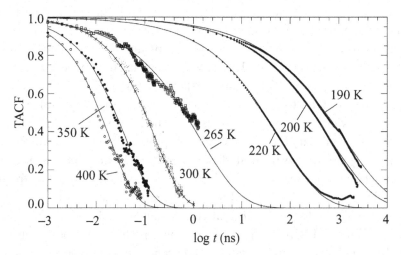

Figure 8.16 Typical TACFs for poly(ethylene). The solid curves are KWW function fits to the data. Data from Boyd *et al.* [21] and Jin and Boyd [39].

where $\phi(t)$ is the value of a torsional angle at time, t, $\phi(0)$ is its value at $t = 0$ and the brackets denote ensemble (system) averages. Example TACFs are shown in Figure 8.16 [21,39]. The solid curves are fits to the data with the KWW function

$$\text{TACF}(t) = \exp - (t/\tau)^{\beta}. \tag{8.2}$$

That the MD generated ACFs for poly(ethylene) have some experimental validity is demonstrated as follows. There are considerable dielectric relaxation data available for dipole decorated semi-crystalline poly(ethylene) (Sec. 9.3.2). For use in comparison with these data, an MD dipolar autocorrelation function (DACF) may be defined in terms of the system moment, $\mathbf{M}(t)$, as

$$\mathrm{DACF}(t) = \langle \mathbf{M}(t) \cdot \mathbf{M}(0) \rangle / \langle \mathbf{M}^2(0) \rangle, \tag{8.3}$$

where the brackets denote a system average and where the normalization is simpler than in eq. (8.1) since the average system moment vanishes in a random distribution of individual orientations. Since the actual PE experiments involve very dilutely dipole decorated polymers, no correlation exists except for self-correlation. In simulations, this allows the placing of virtual dipoles, $\boldsymbol{\mu}$, at each carbon atom (in the plane comprising the atom and each neighbor atom and bisecting the valence angle formed by the trio). Under these conditions eq. (8.3) reduces to

$$\mathrm{DACF}(t) = \langle \boldsymbol{\mu}(t) \cdot \boldsymbol{\mu}(0) \rangle / \mu^2 \tag{8.4}$$

and the magnitude of the dipole, μ, can for convenience be taken as unity. It turns out in poly(ethylene) DACFs, and TACFs are very similar [39]. This is not surprising since the torsional angle conformational jumps drive the dipole reorientation.

Comparison of MD results with frequency domain dielectric data is made by fitting the KWW function to the MD DACFs and locating the frequency domain f_{max} via Fourier transformation. The results are shown in Figure 8.17. The experimental results labeled "α" refer to the crystal phase process in poly(ethylene) (Sec. 9.3.2) and are not relevant to the present discussion. The results labeled "β" are for the amorphous phase primary transition which very probably occurs at lower frequency than in a wholly amorphous poly(ethylene) or, conversely, the β-process in a wholly amorphous poly(ethylene) would likely lie several decades to higher frequency (see, e.g., Figure 9.11). The dashed line indicates schematically this likely shift. The "γ-process" results are for the subglass secondary relaxation and, based on experience (Figure 9.13), probably are not seriously perturbed by the presence of the crystal phase and thus are representative of amorphous poly(ethylene). It is apparent that, experimentally, the primary and secondary processes will have merged in the higher temperature, higher frequency part of the MD results. It is also apparent that the MD results at lower temperature and lower frequency track the upper part of the experimental secondary "γ-process" well. Thus the MD results span the bifurcation region. However, there is little direct evidence for the bifurcation in the ACFs themselves. Curves such as those in Figure 8.16 are well fit by a single KWW function. In part this is due to the fact that experimentally the secondary γ-process is considerably stronger than the amorphous phase primary β-transition (in a ratio of about 3 to 1, see the ε_r, ε_u envelopes in Figure 9.18(a)), and in part due to the considerable width of both processes.

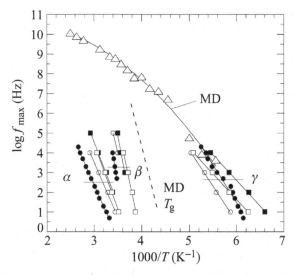

Figure 8.17 Comparison of MD simulations for amorphous poly(ethylene) with experimental dielectric results for semi-crystalline liquid poly(ethylene) (LPE) via a loss peak map. A complete poly(ethylene) dielectric loss map, $\log f_{max}$ vs. $1000/T$ (K) from experimental isochronal temperature scans for the α-, β- and γ-relaxation processes is shown. The β-process is the amorphous phase primary transition and the γ-process is the amorphous phase secondary relaxation. The dashed line schematically indicates the likely shift in the β-process in a completely amorphous polymer. See text for further explanation. The filled points are for lightly oxidized poly(ethylene), the open points are for chlorinated poly(ethylene) (at higher dipole concentration than oxidized specimens). The circles are for LPE and the squares are for branched poly(ethylene). The open triangles are values of $\log f_{max}$ in frequency transformed MD DACFs. The MD volumetric glass transition is indicated by the tick on the temperature axis. Data from Jin and Boyd [39].

Conformational transitions and relaxation

Relaxation processes in polymers must involve conformational changes within the chains. To understand the nature and behavior of these conformational transitions, it is interesting and important to compare the bulk state with that of an isolated and/or phantom chain (one that has no interactions with other chains or longer range interactions with itself). Figure 8.18 compares, via an Arrhenius plot, the conformational transition rates in bulk with those in an isolated chain. The rates in bulk have essentially the same activation energy and similar absolute values. This result is quite curious in that intuitively it might be expected that the bulk rates would be considerably slower and have higher activation energy. Furthermore, the temperature dependence is Arrhenius in nature and the activation energies are very close to the single barrier crossing values [21–28]. This result runs foul of the long standing conundrum that rotation about one bond in the chain independent of other

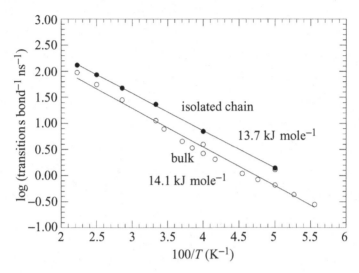

Figure 8.18 Conformational transition rates in bulk poly(ethylene) compared with those in an isolated chain. Data of Boyd *et al.* [21].

bonds would cause large excursions of the attached chain segments in a matrix of intertwined chains and thus would not be possible. Further, the temperature dependence of relaxational phenomena in polymeric melts is invariably non-Arrhenius but rather is VF in nature. The resolution of these questions follows and leads to considerable insight into the nature of glass formation in polymeric melts.

The conundrum of the single-barrier activation energy has been resolved through Brownian dynamics [40,41,42] and MD simulations [43] on isolated poly(ethylene) chains and MD simulations on bulk poly(ethylene) [22–28], where it is found that correlated conformational transitions are common. That is, there are certain conformational sequences in a chain that are favorable to a certain bond undergoing a rotational jump followed by a jump at a nearby bond in the chain, which together leave the rest of the chain relatively undisturbed. The presence of correlation is evident in Figure 8.19, where the probability of a nearby bond jumping after a jump at a given location has taken place is plotted. Results for both the isolated chain case and for bulk poly(ethylene) are shown. It is seen that "next neighbor" correlation, where a bond once removed (position 2 relative to position 0) from the jumping bond has enhanced likelihood of jumping, is present. The Figure also shows that "self-correlation" is present in the bulk polymer, but not in the isolated chains. By "self-correlation" it is meant that there is a finite probability that after a jump in a given bond, it will jump again before other nearby bonds have jumped. Self-correlation is not only present but becomes increasingly common as temperature *decreases*. This observation is important and will be further pursued below.

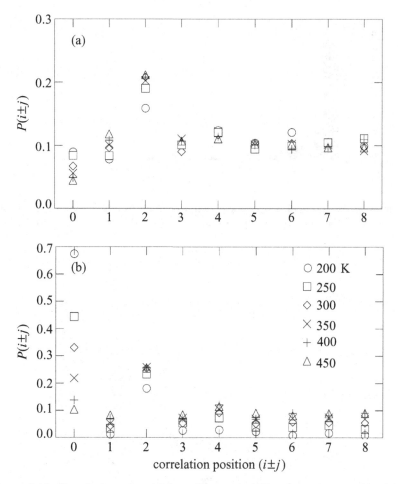

Figure 8.19 Correlation of transitions. The probability that a jump at bond i is followed next by a jump at $\pm j$, where j is restricted to a window of ± 8 bonds. Both isolated chain (a) and bulk (b) results are shown. Data of Boyd *et al.* [21].

Although an element of time has not been introduced into the method of recording correlated events employed in Figure 8.19 it is found that the correlated second jumps occur quickly after the first, as seen in Figure 8.20. However, the correlated jumps are not simultaneous. This feature is central to the single-barrier nature found for the activation energy for conformational transition rates in that it avoids the double-barrier problem. In fact, as seen in Figure 8.21, at the moment of barrier crossing by the first (0) bond, the torsional angle of the second (± 2) bond is typically displaced well away from its barrier top position and the most likely position is near its barrier minimum.

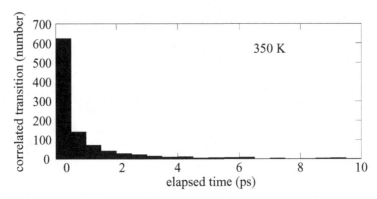

Figure 8.20 The number of ± 2 correlated transitions as a function of the elapsed time between the first bond jump and the second bond jump. Data of Boyd *et al.* [21].

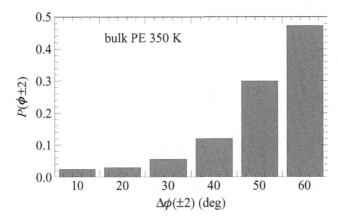

Figure 8.21 The non-simultaneous character of correlated transitions. The probability of the ± 2 bond being displaced by $\Delta\phi$ from its own barrier top while the first bond is crossing the top of its barrier. Data of Boyd *et al.* [21].

It is interesting to see in detail the kinds of conformational sequences that lead to the next-neighbor correlations. These can be identified in simulations by recording the conformation sequence before and after such a correlated event occurs. The common ones that have been proposed and identified are illustrated in Figure 8.22 and can be described as *gauche* migration [44], opposite sign *gauche* pair creation [40–44], kink creation [45], kink inversion, [46,47] and same sign *gauche* pair creation [21]. All of them meet the expected criterion that they lead to minimal disturbance of the chain segments leading away from them.

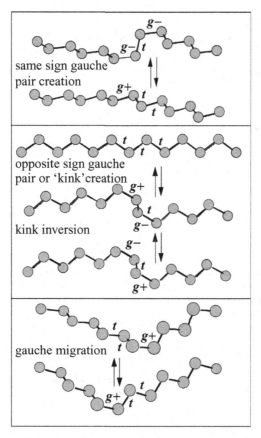

Figure 8.22 Common next-neighbor correlated conformational transitions found in poly(ethylene).

Bulk vs. isolated chain

Relaxation times from KWW function fits at a number of temperatures for both the bulk state and the isolated chain are displayed in Figure 8.23. It is apparent that, as opposed to conformational transition rates (Figure 8.18), the relaxation times do indeed show a wide divergence between bulk and isolated chains. The isolated chain continues to show Arrhenius behavior with single-barrier activation energy to low temperature, while the bulk chain relaxation times show a much stronger VF like temperature dependence. The KWW β width parameters, shown in Figure 8.24, are also quite different. The isolated chain shows near SRT behavior over a wide range of temperatures, whereas the bulk state exhibits broadening as temperature decreases.

The resolution of the question of why the temperature dependences of conformational transition rates vs. relaxation times are so different lies in observing exactly

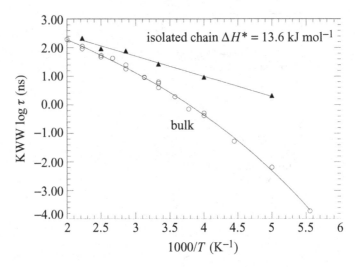

Figure 8.23 Temperature dependence of KWW relaxation times, τ, for the TACF both in bulk and for the isolated chain. The curve through the bulk state results is a VF equation (eq. (6.9)) fit with parameters $A = 3.98$, $B = 702$, $T_\infty = 88.5$. Results of Boyd *et al.* [21].

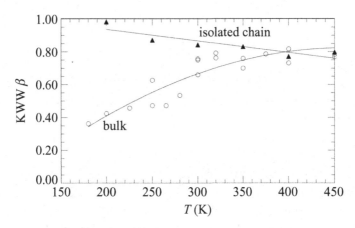

Figure 8.24 KWW equation β parameters for the same temperatures as in Figure 8.23. Results of Boyd *et al.* [21].

where the transitions actually occur with respect to location in the polymer chains. In Figure 8.25 the number of transitions occurring at each bond in a simulation system is shown at four temperatures. At each temperature, the MD trajectories are such that the total number of transitions is the same, 4000. The temperature range spans that of a high temperature melt down to that of the volumetric MD glass transition. It can be seen that there is a great deal of variation in the number of

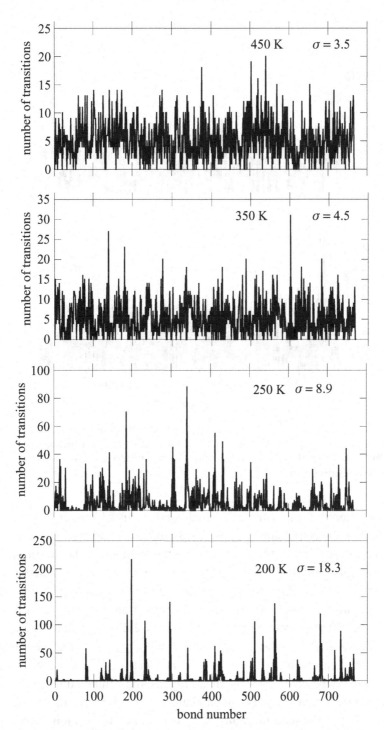

Figure 8.25 Spatial distribution of conformational transitions in bulk poly-(ethylene). The bonds are numbered serially along the chain and the number of transitions at each bond is plotted against the bond number. The standard deviation, σ, of the actual numbers at each bond about the system average $X =$ total transitions/total number of bonds ($X =$ 4000 transitions/765 bonds at all four temperatures $= 5.23$) is shown also. Results of Boyd *et al.* [21].

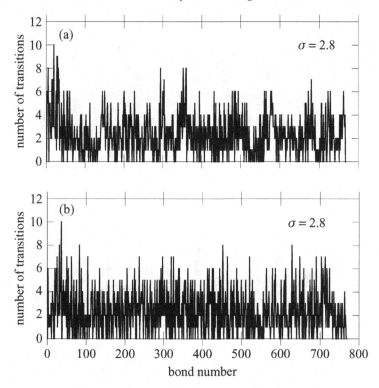

Figure 8.26 Spatial distribution of conformational transitions for isolated chain poly(ethylene) at (a) 450 K and (b) 200 K. The numbers of transitions and bonds and the definition of the standard deviations, σ, are the same as those in Figure 8.25. Results of Boyd *et al.* [21].

transitions occurring among the individual bonds. This becomes especially notice-able at the lowest temperature. A possible source for this variability might be the statistical effect of dealing with a relatively small average number of transitions per bond (4000 transitions/765 bonds = 5.23 in the present example). Statistically, if all bonds were equivalent, this situation would lead to a Poisson distribution where the standard deviation, σ_P, about the average number of transitions per bond, X, is given by $X^{1/2}$ [48]. In the present case $\sigma_P = 5.23^{1/2} = 2.28$. Reference to Figure 8.25 indicates that at 450 K the actual standard deviation, σ, is somewhat larger than σ_P, but more importantly that the discrepancy between the observed and Pois-son standard deviations increases markedly as the temperature decreases, the observed value becoming much larger than the Poisson one. In contrast, in the isolated chain case as seen in Figure 8.26, the standard deviation found in simu-lation is not far from the Poisson value and is temperature independent. Thus it appears that in the bulk the distribution of transition rates over the bonds is not random.

The conclusion that the distribution of transitions over the bonds is non-random in nature is given further substance by considering the time evolution of transitions [49]. Consider a possible extreme situation at low temperatures in the glass where the only transitions taking place over a long time period are associated with certain local conformational sequences such as the ±2 correlated events (Figure 8.19) and the self-correlated over-and-back motions described above and where the locations of these sequences in the polymer chain are fixed for long periods of time. Then the number of transitions at a given "active site" formed by such a fixed sequence should accumulate linearly in time, just as in any zero order reaction. Thus the average number of transitions per bond, X, should also increase linearly in time. The standard deviation, σ, about the average would have a linear time dependence as well. In contrast, the standard deviation, σ_P, of the Poisson distribution for a population of randomly distributed transitions over *all* the bonds, given by $X^{1/2}$, should evolve as $t^{1/2}$. In Figure 8.27(a) it is seen that indeed the transitions and the standard deviation evolve linearly in time over a reasonably long trajectory (25 ns). It is also seen that the Poisson standard deviation, $\sigma_P = X^{1/2}$ is far from the measured standard deviation and, of course, does not follow the observed linear time dependence.

The result in Figure 8.27(a) is found at a temperature (180 K) considerably below the MD T_g (220 K) and over a trajectory whose length is far short of encompassing the entire subglass process. Figure 8.27(b) shows the effect of temperature on the ratio of the MD-measured standard deviation to the Poisson value determined over the same trajectory lengths. The disparity between the two measures is seen to decrease with increasing temperature. Figure 8.28 shows in histogram form the effect of temperature on the distribution of transitions. At the lower temperature of 180 K, well below the MD T_g (= 220 K), the distribution is nearly monotonic with many bonds not jumping at all during the trajectory (25 ns). At the MD T_g the distribution is more symmetrical about the mean number of jumps per bond and all of the bonds have jumped.

The source of the variability in the distribution of transitions over the bonds thus is not to be associated with the vagaries of small numbers. Rather it is more fundamental in nature. The dynamic heterogeneity is the result of the appearance of non-ergodicity over increasingly longer time scales as temperature is lowered. That is, in an equilibrium melt all of the bonds of the same chemical type are equivalent. But in an instantaneous snapshot of the system, of course, the individual bonds are found in many diverse conformational states. As time evolves, each local bond environment experiences a set of states representative of equilibrium (i.e., an ergodic system). In a high temperature melt, the time required for equilibration is short, but lengthens as temperature decreases. Because of the constrictions imposed by the bulk packing environment and the connected nature of chain segments, the

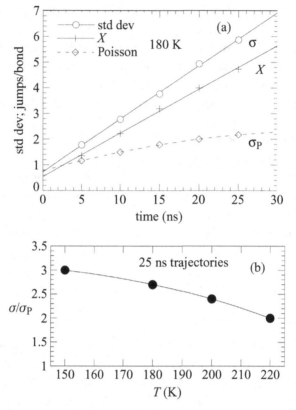

Figure 8.27 (a) Time evolution of dynamic heterogeneity. The average number
of conformational jumps per bond, X, and the standard deviation about the mean,
σ, are plotted against time. Also shown is the standard deviation, σ_P $(= X^{1/2})$,
from a Poisson (random) distribution of the jumps over the bonds. The solid lines
are linear regressions. (b) Effect of temperature on the ratio, σ/σ_P, of the MD
determined standard deviation of jumps per bond to the Poisson (random) value.
Results of Jin and Boyd [49].

time requirement for a given bond to visit its own set of conformational states
is highly dependent on the conformational states of its neighboring bonds. For
example, a bond finding itself in a favorable ± 2 next-neighbor pair sequence
(Figure 8.22) will experience many more transitions than a bond not so favorably
located. Thus at all temperatures there will be a distribution of local transition rates
depending on local conformational circumstances and packing environments. As
the temperature is lowered the sensitivity to these circumstances increases. In fact in
bulk poly(ethylene), at temperatures well below the MD volumetric T_g, essentially
all of the observed TACF relaxation in a reasonably long trajectory (100 ns) is due
to ± 2 next-neighbor pair correlated transitions [39]. This is in contrast to the higher
temperature melt where such transitions are common but not dominant. Since the

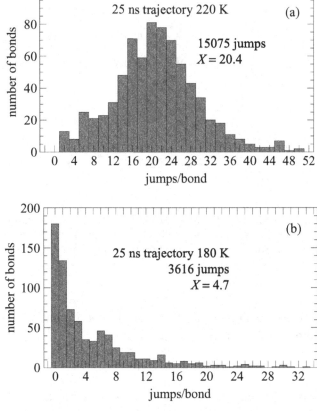

Figure 8.28 The effect of temperature on the distribution of jumps/bond in histogram form: (a) at the MD T_g where all of the bonds have jumped over the MD time trajectory; (b) well below the MD T_g where many of the bonds have not jumped over the same trajectory length.

TACF is a *system* average, its complete decay requires the ergodic participation of *all* bonds. Thus the VF behavior of the KWW relaxation times of the ACF is due to the developing dynamic spatial heterogeneity of the system as the temperature is lowered. Some bonds relax completely (i.e., visit all of their conformational space) and do so rapidly but some bonds do this much more slowly and, well into the subglass region, while, some bonds may experience no conformational transitions at all over the time scale of an experiment.

A simple model demonstrating the relationship between conformational transition rates and segmental autocorrelation functions

The relationship between the distribution of conformational transition rates, which becomes increasing heterogeneous with decreasing temperature, and the

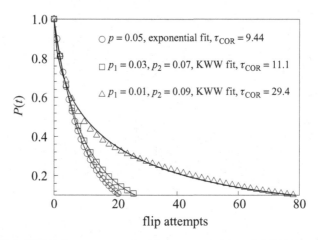

Figure 8.29 ACF $P(t)$ for an uncorrelated two-state spin model. Lines are exponential (homogeneous dynamics) and stretched-exponential (heterogeneous dynamics) fits to the simulation data. Integrated relaxation (correlation times) τ_{COR} for each set of flip rates are also given.

ensemble-average (segmental) ACFs that relax by these transitions can be illustrated with a simple model. Figure 8.29 shows the ACF

$$P(t) = \langle s_i(t)s_i(0)\rangle \tag{8.5}$$

for an ensemble of uncorrelated spins ($s_i = +1$ or -1). In the three cases illustrated, the mean probability p that a given "dihedral" (spin) undergoes a "transition" (a spin flip via a Monte Carlo move) per unit time is constant at $p = 1/20$. In the first case, each dihedral undergoes transitions with the mean probability (dynamically homogeneous), yielding single-exponential decay of the ACF and a correlation time corresponding to the mean transition time. In the second and third cases, half of the dihedrals flip faster and half flip slower than the average (which is unchanged). As anticipated, the greater the dispersion in the transition rates, the longer the (integrated) autocorrelation time. Also note that greater dispersion in the underlying transition rates leads to an ACF with more stretched-exponential character.

Homogenization of conformational transitions

At temperatures and times where relaxation is complete, it is expected that polymers will display *homogeneous* behavior. Specifically, all of the (chemically equivalent) bonds will display the same average transition rates and dynamic heterogeneity will disappear when the system is monitored for sufficient time. The time scale over which the fluctuations in dynamics relax, i.e., the time scale over which dynamical

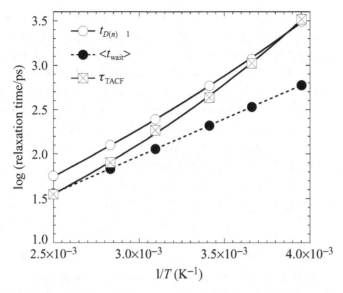

Figure 8.30 Comparison of waiting, homogenization, and the integrated torsional autocorrelation times for alkyl dihedrals in PBD from MD simulations. Data from Smith *et al.* [50].

behavior must be monitored and averaged so that each individual dihedral exhibits "mean" behavior, is referred to as the "homogenization" time scale.

It is possible to use MD simulations to probe the homogenization time scale. For this purpose the distribution of waiting times between transitions for different dihedrals can be monitored as a function of n (the number of transitions) and characterized by their normalized dispersion $D(n) = \sigma^2(n, T)/n^2$, where σ is the second central moment of the distribution. For the PBD melt described in Sec. 6.4, $D(n)$ was observed to decrease with increasing n at all temperatures, i.e., the (normalized) distribution of waiting times becomes relatively narrower with increasing n [50]. This indicates that the conformational dynamics are becoming more homogeneous when averaged over longer time (which increases linearly with n), i.e., that initially fast jumping dihedrals eventually become slow for a period, and initially slow dihedrals become faster. For a given value of n, $D(n)$ was found to increase dramatically with decreasing temperature. It is possible to identify for each temperature the value of n at which $D(n) = 1$ increases with decreasing temperature. For a system with homogeneous conformational dynamics $D(n) = 1$ occurs at $n = 1$, i.e., the normalized dispersion in the waiting time distribution reaches unity on the time scale of a single transition given by $t_{D(n)=1} = n$ and $\langle t_{\text{wait}} \rangle = (1)\langle t_{\text{wait}} \rangle = \langle t_{\text{wait}} \rangle$, where $\langle t_{\text{wait}} \rangle$ is the mean waiting time. The values $t_{D(n)=1} = n$ and $\langle t_{\text{wait}} \rangle$ for the PBD melt are shown in Figure 8.30 as a function of inverse temperature. The "homogenization time" given by $t_{D(n)=1}$ shows a much stronger temperature dependence than $\langle t_{wait} \rangle$ and is very similar to the segmental relaxation time as monitored by the torsional autocorrelation time in both magnitude and temperature dependence. Moreover, this correspondence demonstrates that the non-Arrhenius temperature dependence of the microscopic conformational relaxation (TACF), and hence segmental relaxation in the polymer melt, as T_g is approached, is strongly tied to the increasingly heterogeneous nature of conformational dynamics with decreasing temperature. Furthermore, the segmental relaxation appears to reflect the time scale for homogenization of conformational dynamics, i.e., the time scale on which most dihedrals undergo something like the mean number of transitions. These results are consistent with multidimensional NMR studies (Sec. 8.1.2) that have indicated homogenization of segmental dynamics occurs on the α-relaxation time.

Kinetic description of the source of dynamic heterogeneity

The curious observation that the overall *system* conformational transition rate remains Arrhenius in nature while the relaxation time behavior diverges as temperature decreases also needs to be addressed. The explanation also lies in the effects of dynamic spatial heterogeneity. As temperature decreases, the system transition rate becomes more dominated by the self-correlated, "over-and-back" ± 0 transitions (Figure 8.19). Yet the number of sites where they occur is limited (see Figure 8.25, 200 K) and so they are ineffective in relaxing the system TACF. So the questions are, why do these sites have such high individual transition rates and why are they concentrated a relatively few places? In turn these are connected to lifetime issues.

In thinking of conformational jumps as thermally activated barrier crossings it is natural to consider a jump as taking place from the vicinity of a local minimum over the rotational barrier (plus whatever contribution from the packing environment) to the vicinity of another minimum. In both positions higher frequency torsional oscillations about the minima take place. At high temperature, indeed this is the case. However, as the temperature is lowered the constrictive effects of the surrounding packing result in the time spent in positions well away from the rotational minima increasing greatly. This is established by monitoring in simulation the average residence time at a given torsional angle and comparing it with that expected in an ergodic system. An *effective potential* $V_{eff}(\phi)$ can be determined by sampling the torsional state of each bond along a trajectory. The sampling produces a probability function, $p(\phi)$, for finding a bond in torsional state, ϕ, that in an ergodic system is a Boltzmann probability in terms of the effective potential, $V_{eff}(\phi)$, and the partition function, Q:

$$p(\phi) = \exp -(V_{eff}(\phi)/kT)/Q, \tag{8.6}$$

which on inversion gives

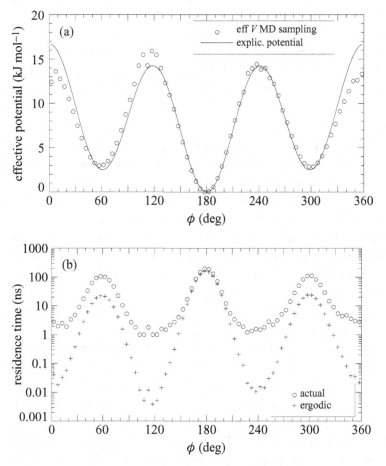

Figure 8.31 Ergodicity of bond populations: bulk poly(ethylene) at 180 K. (a) The effective torsional potential derived from torsional angle sampling of a 100 ns trajectory (circles). The explicit potential used in the simulation is super-posed as the solid curve. (b) The average residence time (in a torsional angle interval) derived from sampling (circles). The crosses are the residence times that would result under the ergodic condition that all bonds visit each torsional angle interval equally. Results of Y. Jin and Boyd [39].

$$V_{\text{eff}}(\phi) = -kT(\ln(p(\phi)) - \ln Q). \tag{8.7}$$

An effective potential, with its minimum adjusted to zero, for bulk poly(ethylene) at 180 K (well into the glassy region) determined from a 100 ns trajectory by sampling at 0.1 ns intervals is shown in Figure 8.31. The explicit torsional potential used in the simulation is overlaid on the effective potential. The two sets of data agree fairly well but the effective barrier from sampling at $\phi = 0°$ is noticeably lower than the explicit one. This, in turn, implies some overpopulation frozen-in from a higher temperature and not equilibrated in the glass.

Average residence times in the torsional states can also be determined by sampling in the MD simulation [39]. An average residence time per bond at each $\Delta\phi$ sampling interval for the bond population can be computed by recording the number of bonds that were sampled in each $\Delta\phi$ interval over the complete (100 ns) trajectory. The time spent by all the bonds in a given $\Delta\phi$ interval is the number of sampled points in the $\Delta\phi$ interval over the entire sampling time trajectory multiplied by the time sampling interval (0.1 ns). The average residence time is then the total time spent in a given $\Delta\phi$ interval divided by the number of bonds found to have visited the interval. If the bonds were equivalent (ergodic system), all of the bonds would have visited each $\Delta\phi$ interval equally over a long trajectory. The average residence time is then the total sampled time occupancy of each $\Delta\phi$ interval divided by the number of bonds in the system. Figure 8.31 also compares the actual per bond average residence times with the values expected on the basis of bond equivalency. It may be seen that although the two measures approach each other at the more populated positions, the actual residence times at higher energy positions are far longer than on the equivalence basis.

In fact, the residence times are so long that bonds can be considered to be trapped by packing constraints long enough that torsional oscillations take place about these positions and the torsionally displaced sites become eligible centers for the initiation of a conformational jump. This results in the effective barrier becoming the difference in energy between the trapped position and the barrier top and thus the barrier becomes lessened. This is illustrated schematically in Figure 8.32. In a high temperature melt, which tends to provide theta solvent conditions, the probability of a bond being found at a given torsional angle position is driven by the torsional potential. That is, the effective potential of eq. (8.6) and eq. (8.7) is well represented by the explicit torsional potential used in simulation. The fractional population of bonds at a given torsional angle ϕ is represented by Boltzmann weighting as

$$p(\phi) = \exp(-V(\phi)/kT)/Q, \tag{8.8}$$

where $V(\phi)$ is the torsional potential and Q is the classical partition function. Although this equilibrium expression presumably will not strictly apply in the glass (i.e., the overpopulation apparent at $\phi = 0°$ in Figure 8.31) it is instructive to invoke it as an approximation. Referring again to Figure 8.32, the flux of transitions with activation energy $\Delta E_{\text{Tr}}^* = \Delta E^* - V(\phi)$ will be proportional to

$$p(\phi)\exp(-\Delta E_{\text{Tr}}^*/kT) = p(\phi)\exp(-(\Delta E^* - V(\phi))/kT) = Q^{-1}\exp(-\Delta E^*/kT). \tag{8.9}$$

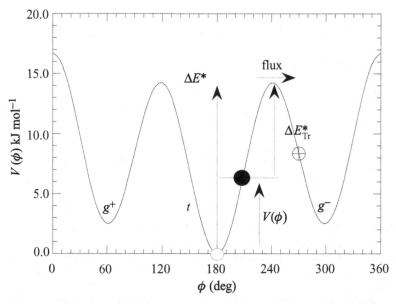

Figure 8.32 A schematic rendering of the effect of long torsional angle residence times on transitional behavior. At high temperature, bonds may be considered to oscillate about a minimum in the torsional potential (open circle at 180°). The activation energy for a barrier crossing, ΔE^*, is the full barrier height. However, at low temperature a bond can occupy a position far from the torsional barrier minimum (filled circle at $V(\phi)$) for a time long enough for it to be considered as the starting point for an activated jump to another long-lived site (at the circled cross). The activation energy then is the difference between the full barrier height and the position on the torsional potential $\Delta E_{\mathrm{Tr}}^* = \Delta E^* - V(\phi)$.

The partition function integral, Q, has only minor temperature dependence compared to the exponential activation energy term. Thus the reduced barrier and the decreased population from Boltzmann weighting are compensating effects and the *system* flux (i.e., the total transition rate) is similar for the trapped bond case in comparison with that for the free bonds. However, the rates at *individual* bonds will be quite different. The rate for a given bond at the filled circle position in Figure 8.32 will be much higher than for a bond at the open circle and the former will show up as a large spike in plots like Figure 8.25. But conversely, there would not be many of these spikes due to the Boltzmann weighting effect. They thus would not contribute significantly to the system ACF decay. So, in essence, the maintenance of Arrhenius temperature dependence of transition rates is a consequence of two opposing effects in the dynamic heterogeneity that accompanies lowering temperature.

An example of a trajectory of a bond where self-correlated ±0 transitions occur at a rapid rate from a location on the torsional potential significantly away from

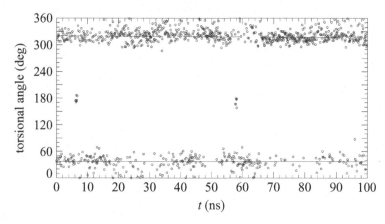

Figure 8.33 Time trajectory of repeated ±0 self-correlated over-and-back transitions $g^+ \Leftrightarrow g^-$ at a given bond in a poly(ethylene) MD simulation at 180 K. The points are sampled values of the torsional angle at time intervals along the trajectory. The lines are averages for each position over the trajectory. Both positions are noticeably distorted toward the $\phi = 0°$, 360° torsional potential barrier thus decreasing the effective transition barrier. Data of Y. Jin and Boyd [39].

the minimum to another such site, like that described in Figure 8.32, is shown in Figure 8.33. It can also be seen that the density of sampled points is not the same for each state. The near 300° state is somewhat more populated on average than the near 60° state. In terms of Figure 8.32 this is due to an energy difference between the local 'ground states' of the filled circle and the circled cross and consequent Boltzmann weighting effects (Appendix AII, eg. (AII.15)).

Role of the surrounding packing

To this point, the role of the torsional potential in driving conformational events has been emphasized. It is also important to assess the role of the surroundings in which the events take place. Since the individual events, i.e., the conformational transitions, are activated jumps it is of interest to see to what degree the hindrances due to the surrounding packing contribute to the associated activation barriers. This can be accomplished by locating individual jump events along the MD trajectory and computing the energy of interaction of a jumping segment with its surrounding chains during the jump event. This latter energy is then included with the intramolecular energy of the jumping segment in establishing the energy profile of the event. Figure 8.34 illustrates such a profile. It may be seen that the barrier U_1 ($\sim$20 kJ mol^{-1}) is substantially higher than the explicit torsional barrier (14 kJ mol^{-1}). It is also to be noticed that there is an energy difference between the "before" and "after" ground states ($U_1 \neq U_2$). From the torsional angle trajectory

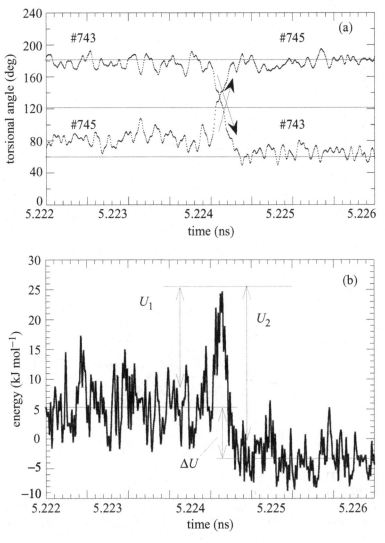

Figure 8.34 Energy profile for a ±2 correlated *gauche* migration event in poly(ethylene) at 180 K: (a) the torsional angle trajectory at bonds numbered 743 and 745; (b) the accompanying energy profile. U_1, U_2 denote the energy of the barrier relative to the *average* energies (denoted by horizontal lines) before and after, respectively, the event. The difference in average energies, before and after, is denoted by ΔU. Data of Boyd and W. Jin [51].

this can be seen to be due to bond 745 being displaced from the explicit torsional energy minimum (at 60°) before the jump but arriving at the minimum afterwards (at 180°). The presence of such ground state energy differences is common, as may be seen in Figure 8.35. They are important because they are sufficient to influence, through Boltzmann weighting, the strength of the relaxation process

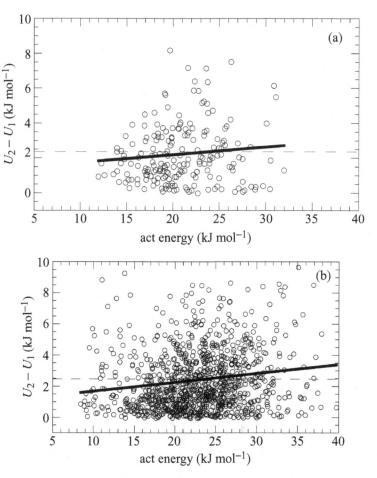

Figure 8.35 The differences in ground state energies $U_2 - U_1$ (see Figure 8.32) for the ± 2 correlated conformational jumps occurring over 250 ns trajectories in poly(ethylene) at (a) 180 K and (b) 220 K, plotted against the jump activation energy (U_1 in Figure 8.32). The bold line is a linear regression and the dashed line is the mean value. Data of Boyd and W. Jin [51].

(Appendix AII). That is, the strength contribution of an individual site will decrease (see eq. (AII.15)) as temperature decreases and therefore the overall relaxation strength will decrease as well.

The results of analyses such as those in Figure 8.34 for all of the ± 2 correlated transitions occurring over 250 ns trajectories in poly(ethylene) at 180 K, 200 K and 220 K are summarized in the form of histograms in Figure 8.36. It can be seen that the explicit torsional barrier invoked in simulation tends to mark the onset of activation energy values with most of the transitions having significantly higher values. It is also to be remarked that only at 220 K is the relaxation complete over

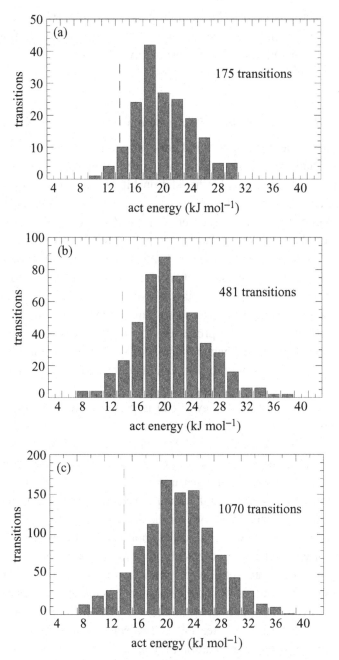

Figure 8.36 Distribution of activation energies for conformational transitions in poly(ethylene) at (a) 180 K, (b) 200 K and (c) 220 K. The vertical dashed line indicates the value of the explicit torsional barrier. Data of Boyd and W. Jin [51].

the 250 ns trajectories used in the sampling (see Figure 8.16). Thus only at that temperature can the histogram be considered to be complete. At the two lower temperatures the histograms would presumably add higher values over a longer trajectory.

8.2.2 Bifurcation of the primary and secondary processes from MD simulations

Experimentalists, theorists, and simulators have long sought to understand the molecular mechanisms of the main chain secondary (subglass) β-relaxation in amorphous polymers, in particular the relationship of the β-relaxation to the fundamental molecular motions available to polymers, i.e., conformational librations and transitions. While it is generally agreed that the β-process involves local motion of the polymer chain backbone leading to partial relaxation of the polymer, the nature of these motions, whether they are spatially and temporally homogeneous or heterogeneous, and the relationship between molecular motions responsible for the β- and α-processes remain controversial. MD simulations utilizing validated potentials have been demonstrated to accurately reproduce the structural and dynamic properties, including relaxation processes, for a wide variety of polymers [52], including those discussed in Chapter 6 and Sec. 8.2.1. Hence, atomistic MD simulations appear to be well suited to help elucidate the molecular mechanism of the β-relaxation and how it relates to the α-process.

From a simulator's point of view, PBD is perhaps the most appealing glass-forming polymer due to its simple chemical structure, lack of bulky side groups and non-polar nature. Consequently, PBD can be simulated more effectively than perhaps any other glass-forming polymer for which extensive experimental data are available. Despite these advantages, however, resolving the α- and β-relaxation even in PBD remains a major computational challenge, as illustrated in Figure 8.37, which reveals that the relaxation times for the dielectric α- and β-processes in PBD [11,53] are well resolved only on time scales approaching milliseconds. Hence, simulation trajectories of tens or hundreds of microseconds are required to adequately separate polymer segmental motions responsible for the α- and β-relaxations and allow definitive assignment of molecular mechanisms to these processes. As shown in Figure 8.37, even after years of simulation of PBD melts (Sec. 6.4) utilizing a quantum-chemistry-based UA potential [54], trajectories are not of sufficient length to fully resolve the α- and β-relaxation processes. The same situation prevails for amorphous poly(ethylene), as illustrated in Figure 8.16.

Low barrier PBD

Dielectric spectroscopy experiments on glass forming liquids under high pressure [55] have shown that increasing pressure can result in a large shift in the relaxation

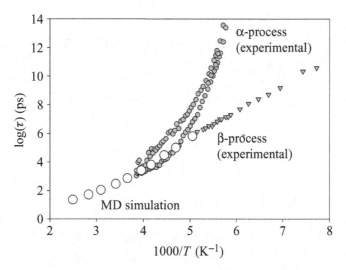

Figure 8.37 Relaxation times for the dielectric α-relaxation process (small circles) and dielectric β-relaxation process (triangles) obtained from experimental measurements on melts of PBD as a function of inverse temperature from Arbe *et al.* [11] and Aouadi *et al.* [53]. Also shown are relaxation times for the (apparent) combined relaxation process obtained from MD simulations (large circles) from Smith and Bedrov [54]. The tendency of the relaxation time for the combined process as obtained from MD simulations to follow the experimental β-relaxation process at the lowest simulated temperature is due to the inability of the simulations to adequately sample the slow α-relaxation process.

time of the cooperative α-process relative to that of the local β-process which is relatively insensitive to pressure. As a result, the bifurcation of the relaxation processes is moved into a shorter time window at higher temperature. In MD simulations the effect of raising pressure can be mimicked by reducing intramolecular energy barriers for dihedral transitions of the polymer backbone. In such an exercise, the intramolecular energetic effects are reduced while leaving the intermolecular packing essentially unperturbed, similar to the effect of raising pressure and temperature. Reducing dihedral barriers should therefore move the bifurcation into a shorter time window at lower temperature. To create melts with faster dynamics the PBD model described in Sec. 6.4, which is referred to here as the chemically realistic PBD, or CR-PBD, was modified with all backbone dihedral potentials (excepting double bonds) reduced by a factor of 4 relative to the CR-PBD model, yielding the low barrier or LB-PBD model [54,56]. All other bonded and non-bonded interactions in these models are identical. The reduction of dihedral barriers in the PBD melt allows essentially identical structural properties to those obtained using the CR-PBD model – only dynamic properties are significantly influenced by the removal of the dihedral barriers.

Segmental relaxation in LB-PBD

Segmental relaxation in LB-PBD melts has been probed by monitoring the TACF, given by eq. (5.3), for $C_{sp^2}-C_{sp^3}-C_{sp^3}-C_{sp^2}$ (alkyl) and $C_{sp^2}-C_{sp^2}-C_{sp^3}-C_{sp^3}$ (allyl) dihedrals, as well as the DACF given by eq. (8.4). The ACFs have been fit as a single relaxation process and as a sum of two processes, labeled β (short-time) and α (long-time)

$$\text{ACF}(t) = A_\beta f_\beta(t) + A_\alpha f_\alpha(t). \tag{8.10}$$

Here $f_\alpha(t)$ and $f_\beta(t)$ are KWW functions representing the α- and β-relaxations, respectively, while A_α and A_β are amplitudes of these processes with the constraint $A_\alpha + A_\beta \leq 1.0$. Figure 8.38 shows the alkyl TACF and DACF for LB-PBD melts over a wide range of temperatures obtained from MD simulations as well as the best representations of the TACF that could be obtained assuming a single relaxation process and a sum of two relaxation processes. It can be clearly seen that a single relaxation process fails to provide an accurate description of the decay of the TACF and DACF, while significant improvement in the description of the ACFs can be obtained when a sum of two relaxation processes is utilized. In Figure 8.39 the dielectric loss obtained from the Fourier transform of the DACF using the DACF obtained from fitting MD simulation results with a single process and a sum of two processes is compared for the LB-PBD melt at 140 K. The presence of two relaxation processes in the LB-PBD melt is even clearer in the frequency domain than in the time domain.

Temperature dependence of the α- and β-relaxation times in LB-PBD

The integrated relaxation times (eq. (6.11)) for the α- and β-relaxation processes obtained from fitting the decay of the TACF for the alkyl and *cis* allyl dihedrals as well as from fitting the decay of the DACF in the LB-PBD melts are shown as a function of inverse temperature in Figure 8.40. As anticipated, lowering the dihedral barriers in the PBD melt has moved the bifurcation of the α- and β-relaxation processes into the time window accessible to MD simulations. At the lower end of the temperature range investigated the separation of the α- and β-relaxation times is greater than four orders of magnitude in time. It was found that the α- and β-relaxation processes in LB-PBD melts do not merge completely even at higher temperatures. The α-relaxation times can be described well by a VF temperature dependence, while the β-relaxation times can be described well by an Arrhenius temperature dependence as shown in Figure 8.40.

Figure 8.40 reveals that the β-relaxation time for the *cis* allyl dihedrals is shorter than the β-relaxation time for the alkyl dihedrals for all temperatures, consistent with the lower conformational energy barriers for rotation about the allyl dihedrals

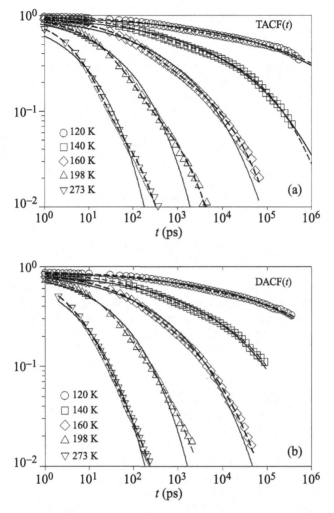

Figure 8.38 The alkyl TACF and DACF for LB-PBD melts at selected temperatures (symbols). Also shown are fits to a single relaxation process (solid lines) and to a sum of two relaxation processes (dashed lines). Data from Smith and Bedrov [54].

compared to the alkyl dihedrals in PBD. At higher temperatures the α-relaxation time for the *cis* allyl dihedrals is also shorter than that for the alkyl dihedrals. However, with decreasing temperature the α-relaxation times for the two types of dihedrals appear to merge. This behavior may reflect the increasingly important role of cooperative matrix motion in segmental relaxation in polymer melts with decreasing temperature. At higher temperatures complete relaxation (i.e., the α-relaxation process) for the *cis* allyl dihedrals may occur more rapidly than that for the alkyl dihedrals due to the lower dihedral barrier for the former that allows

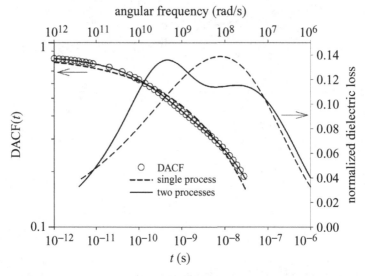

Figure 8.39 The DACF and its representation with a single relaxation process and sum of two KWW relaxation processes as a function of time for the LB-PBD melt at 140 K. Also shown is the corresponding dielectric loss as a function of angular frequency. Data from Smith and Bedrov [54].

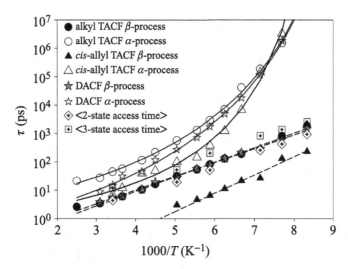

Figure 8.40 Various integrated relaxation times and conformational state access times obtained for LB-PBD melts as a function of inverse temperature. The lines are VF (solid) and Arrhenius (dashed) representations of the temperature dependence of the α- and β-relaxation times, respectively. Data from Smith and Bedrov [54].

relaxation to occur with less cooperative motion of the matrix. At lower temperatures (and higher density) significant cooperative matrix motion is required for the
α-relaxation to occur in either the *cis* allyl or alkyl dihedrals and little difference is
seen in their relaxation times.

Interestingly, while the β-relaxation time for the DACF in the LB-PBD melts
closely follows that of the alkyl dihedrals for all temperatures, the α-relaxation time
for the DACF is systematically shorter than that for the alkyl dihedrals at higher
temperatures. At lower temperatures, where matrix effects begin to dominate relaxation behavior, the α-relaxation time for the DACF merges with that for the alkyl
dihedrals. Dipole relaxation in PBD is due primarily to angular reorientation of
the *cis* H–C=C–H groups that have a net dipole moment (unlike *trans* H–C=C–H
groups, which have no net dipole moment due to symmetry). Therefore, it is reasonable to assume that dipole relaxation in PBD will be influenced by the relaxation
behavior of both the *cis* allyl and alkyl dihedrals. Figure 8.40 clearly reveals that
the β-relaxation process for the DACF is dominated by relaxation (β-process)
of the relatively slowly relaxing alkyl dihedrals. However, the α-relaxation time for
the DACF lies between those for the slowly relaxing alkyl and rapidly relaxing *cis*
allyl dihedrals, indicating that both dihedrals play a role in complete orientational
relaxation of the *cis* H–C=C–H groups, as anticipated. Since the α-relaxation time
for the fast *cis* allyl dihedrals merges with that for the slow alkyl dihedrals at low
temperatures, the α-relaxation time for the DACF also merges with that for the
alkyl dihedrals as temperature is reduced.

Conformational motion and the β-relaxation process in LB-PBD

Alkyl dihedrals in PBD undergo transitions between *gauche*$^+$, *trans* and *gauche*$^-$
conformational states. $P_{3\text{-state}}(t)$ is defined as the probability that an alkyl dihedral
has *not* visited all three conformational states and $P_{2\text{-state}}(t)$ as the probability that
a dihedral has *not* visited both *trans* and *gauche* (*gauche*$^+$ or *gauche*$^-$) states after
time t, respectively. Hence $P_{3\text{-state}}(t)$ decays completely when *all* alkyl dihedrals
have visited all three conformational states and $P_{2\text{-state}}(t)$ decays completely when
all alkyl dihedrals have visited both the *trans* and at least one of the *gauche* states.
$P_{2\text{-state}}(t)$ and $P_{3\text{-state}}(t)$ as well as the integrated α- and β-relaxation times for the
TACF for the alkyl dihedrals ($f_\alpha(t)$ and $f_\beta(t)$ from eq. (8.9)) are shown in Figure 8.41
for the LB-PBD melt at 120 K. The close correspondence between $P_{2\text{-state}}(t)$ and
the β-relaxation process is clear, revealing that large scale conformational motions
(conformational transitions between *trans* and *gauche* states) occur on the time
scale of the β-relaxation. Furthermore, both $P_{2\text{-state}}(t)$ and $P_{3\text{-state}}(t)$ decay almost
completely on time scales greater than the β-relaxation time τ_β, say $5\tau_\beta$, but before
significant α-relaxation occurs. Hence, when the α- and β-relaxation processes are
well separated in time, nearly all dihedrals visit all three conformational states

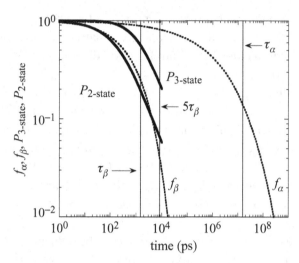

Figure 8.41 The fraction of C_{sp^2}–C_{sp^3}–C_{sp^3}–C_{sp^2} (alkyl) dihedrals that have not accessed two or three conformational states as a function of time for the LB-PBD melt
at 120 K. Also shown are the relaxation functions obtained from fitting the TACF
for the LB-PBD melt at 120 K with a sum of two KWW relaxation processes.
The vertical solid lines denote important time scales. Data from Smith and Bedrov
[54].

before significant α-relaxation occurs. The relationship between the β-relaxation
process and conformational transitions is further illustrated when the average time
needed for alkyl dihedrals to visit both *trans* and *gauche* (*gauche$^+$* or *gauche$^-$*)
states (⟨2-state access time⟩) and the average time needed for alkyl dihedrals to
visit all three conformational states (⟨3-state access time⟩) are compared with the
β-relaxation time scale as shown in Figure 8.40. These times, as well as their
temperature dependence, correspond closely with the β-relaxation time for the
alkyl dihedrals.

Width of the β-relaxation process in LB-PBD

As discussed in Sec. 7.2, the β-relaxation process in glass forming polymers is
typically broad in the time/frequency domain and the width of the process often
increases with decreasing temperature. In the time domain the width of the process
is characterized by the stretching exponent β in the KWW fit. For the TACF for
the alkyl and *cis* allyl dihedrals in LB-PBD melts the KWW β exponent decreases,
i.e., the process broadens, with decreasing temperature [54]. In order to better
understand the origin of the broadening of the β-relaxation process with decreasing temperature in the LB-PBD melts the distribution of times required for alkyl
dihedrals to visit all three conformational states has been monitored. As illustrated

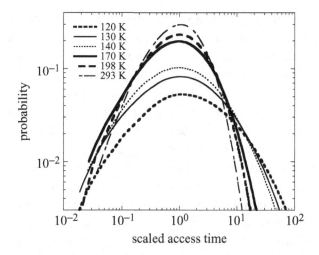

Figure 8.42 The distribution of times required for $C_{sp^2}-C_{sp^3}-C_{sp^3}-C_{sp^2}$ (alkyl) dihedrals to access all three conformational states for LB-PBD melts at various temperatures. The time scale has been normalized by the most probable access time for each temperature. Data from Smith and Bedrov [54].

in Figure 8.40 and Figure 8.41 there is a close correlation between the β-relaxation process and the time scale on which dihedrals visit each conformational state. Figure 8.42 reveals that the distribution of times in which individual alkyl dihedrals visit each conformational state is broad and broadens substantially with decreasing temperature, consistent with the temperature dependence of the stretching exponents for the β-relaxation process in the LB-PBD melts. Figure 8.42 also indicates that the β-relaxation process is quite *heterogeneous* in the sense that individual dihedrals accomplish the exploration of conformational states corresponding to the β-relaxation process at very different rates.

Strength of the β-relaxation process in LB-PBD

The strength of the β-relaxation process in LB-PBD melts is characterized by A_β for the TACF and DACF from eq. (8.10). The β-relaxation in LB-PBD increases significantly in strength with increasing temperature [54,56], becoming the dominant relaxation process at temperatures well above the glass transition temperature. These results are consistent with dielectric measurements on PBD melts [11] that also reveal a dominant β-relaxation process at high temperatures whose strength decreases with decreasing temperature.

In order to better understand the mechanism of the (main chain) β-relaxation process and the temperature dependence of the strength of the β-relaxation process in LB-PBD melts, it is constructive to consider what must happen in order for segmental relaxation, as monitored by the TACF, to occur. The TACF (eq. (5.3)) decays completely only on the time scale over which *all* dihedrals populate each conformational state with near-equilibrium probability. In other words, when the conformation of *individual* dihedrals is monitored over the time scale of the complete decay of the TACF, each dihedral will occupy each conformational state for a fraction of this time equal (or nearly equal) to the equilibrium populations (e.g., 52% *gauche*$^+$ and *gauche*$^-$, 48% *trans* for the alkyl dihedrals of LB-PBD at 140 K) obtained by averaging over all dihedrals. This complete relaxation occurs on time scales longer than the α-relaxation time. On times scales longer than the β-relaxation time, say five β-relaxation times ($5\tau_\beta$), but short compared to the α-relaxation time, partial relaxation corresponding to the β-relaxation process is observed. On this time scale, individual dihedrals do *not* occupy each conformational state for a fraction of this time equal to the equilibrium populations, and the TACF decays only partially.

The contribution of individual alkyl dihedrals to the decay of the TACF in LB-PBD melts after $5\tau_\beta$, corresponding to a time that is sufficiently long for the β-relaxation process to be largely complete and short enough such that significant α-relaxation has not occurred at lower temperatures, has been investigated [54,56]. This was done by monitoring the conformational state of the individual alkyl dihedrals over $5\tau_\beta$ and sorting the dihedrals based upon the fraction of this time spent in the *trans* state (P_{trans}). Figure 8.43 shows the fraction of alkyl dihedrals with P_{trans} smaller than a given value for LB-PBD melts at several temperatures. For example, over a time $5\tau_\beta$ at 140 K, only a small fraction, specifically $0.14 = (0.57 - 0.43)$, of the alkyl dihedrals have P_{trans} near the equilibrium value, specifically $P_{trans} = 0.48 \pm 0.15$, where 0.48 is the equilibrium probability. While most alkyl dihedrals have visited both *trans* and *gauche* states during the $5\tau_\beta$ time window; i.e., $0 < P_{trans} < 1$ for the majority of dihedrals, the distribution of P_{trans} over $5\tau_\beta$ becomes increasingly broad (further from equilibrium) with decreasing temperature. Those dihedrals with $P_{trans} \approx P_{eq}$ over $5\tau_\beta$ will contribute significantly to the decay of the TACF on this time scale (i.e., to the β-relaxation process) and those that spend most of their time in either the *trans* ($P_{trans} \approx 1$) or *gauche* states ($P_{trans} \approx 0$) will contribute little to the β-relaxation process.

For the purpose of investigating the contribution of individual dihedrals to the decay of the TACF the alkyl dihedrals have been grouped into subpopulations whose average P_{trans} (over $5\tau_\beta$) is equal to P_{eq}, i.e., $\langle P_{trans} \rangle_{subpopulation} = P_{eq}$, where the average is taken over P_{trans} of all member dihedrals of the subpopulation. The subpopulations are constructed by taking dihedrals corresponding to the first 2% of the P_{trans} distribution (those dihedrals with the smallest P_{trans}) and adding to this subpopulation the number of dihedrals from the other end of the distribution (largest P_{trans}, or equivalently the smallest P_{gauche}) required to achieve $\langle P_{trans} \rangle_{subpopulation} = P_{eq}$.

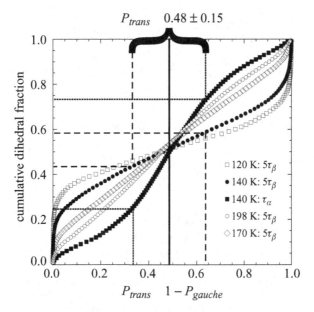

Figure 8.43 The fraction of $C_{sp^2}-C_{sp^3}-C_{sp^3}-C_{sp^2}$ (alkyl) dihedrals that spend a smaller fraction of time in the *trans* conformational state over a period of $5\tau_\beta$ (P_{trans}) than a given value for the LB-PBD melts at various temperatures. Also shown is the distribution obtained over τ_α for the 140 K melt as well as the equilibrium *trans* probability at 140 K (solid vertical line). The dashed and dotted lines illustrate the fraction of dihedrals with near equilibrium populations measured over $5\tau_\beta$ and τ_α, respectively, for the 140 K melt. Data from Smith and Bedrov [54].

These are the dihedrals that are furthest from equilibrium in their occupation of conformational states over $5\tau_\beta$, i.e., the dihedrals that spend the largest fraction of their time in either the *trans* or the *gauche* states. This process is repeated, taking the next 2% of the P_{trans} distribution and adding dihedrals from the other end of the distribution until all dihedrals are divided into subpopulations. When so constructed each subpopulation contains about 4% of the dihedrals.

The TACF after $5\tau_\beta$, or TACF($t=5\tau_\beta$), for each subpopulation of the alkyl dihedrals in the LB-PBD melt is shown in Figure 8.44 for a range of temperatures. The dihedrals on this time scale can be divided into three classes as indicated by the solid lines in the figure. The first class consists of those dihedrals that do not contribute significantly to the decay of TACF($t=5\tau_\beta$), consisting, for example, of 30–35% of the dihedrals at 140 K. The majority of these dihedrals are *not* quiescent since they have visited both *trans* and *gauche* states, be it very asymmetrically, i.e., they spend a large fraction of the $5\tau_\beta$ time window in one state or the other, but not all of it. Only around 4% of the alkyl dihedrals have not visited both *trans* and

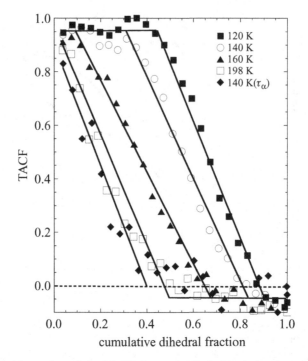

Figure 8.44 The value of the TACF for each alkyl dihedral subpopulation after $5\tau_\beta$ for LB-PBD melts at various temperatures. Also shown is the TACF for each subpopulation after τ_α for the three LB-PBD at 140 K. The solid lines delineate three classes of dihedrals (non-relaxing, partially relaxed and fully relaxed). Data from Smith and Bedrov [54].

gauche states at 140 K after $5\tau_\beta$. The second class consists of dihedrals that have partially relaxed after $5\tau_\beta$ comprising 50–55% of the dihedrals at 140 K. Finally, about 15% of the dihedrals have completely relaxed after $5\tau_\beta$ at 140 K. These dihedrals spend more than 30% of their time in their minority state (i.e., $0.3 < P_{trans} < 0.7$).

The fraction of dihedrals that are partially relaxed on time scales longer than the β-relaxation time but before onset of significant α-relaxation, indicated by the sloped lines in Figure 8.44, is approximately independent of temperature. The decrease in the strength of the β-relaxation process, given as the combined contribution of all subpopulations to the decay of the TACF, with decreasing temperature can be attributed to an increasing fraction of dihedrals that do not contribute to the decay of the TACF and a decreasing fraction of dihedrals that are completely relaxed. Note, however, that nearly all dihedrals, even those that do not contribute significantly to the decay of the TACF, visit both *trans* and *gauche* states (and hence undergo conformational transitions) on the time scale of the β-relaxation process at all temperatures. The β-relaxation process is hence *homogeneous* in the sense that it corresponds to large-scale conformational motions of *all* (or nearly all) dihedrals. A picture of the β-process resulting from motion of a subset of "active" dihedrals that becomes increasingly small with decreasing temperature, leading to a corresponding reduction in relaxation strength, is therefore incorrect, at least for LB-PBD melts. The decreasing strength of the β-relaxation process with decreasing temperature in fact indicates that the complete exploration of conformational space (i.e., every dihedral visiting each conformational state) associated with the β-relaxation process becomes less effectual in fomenting relaxation with decreasing temperature. This results from the increasingly heterogeneous manner in which individual dihedrals visit *trans* and *gauche* states on the time scale of the β-relaxation, manifested in an increasing fraction of dihedrals spending a decreasing fraction of their time (over $5\tau_\beta$) in their minority state.

The relationship between the α- and β-relaxation processes in LB-PBD

The time scale for all dihedrals to simply visit (occupy for some period of time) each conformational state, associated with the β-relaxation process in LB-PBD melts, is much shorter than the time required for every dihedral to achieve equilibrium occupancy of conformational states. As shown in Figure 8.43, the latter occurs on time scales longer than but comparable to the α-relaxation time τ_α. Specifically, Figure 8.43 shows that when the state of the individual dihedrals is averaged over time τ_α a substantial fraction $(0.74 - 0.24 = 0.50)$ of the alkyl dihedrals at 140 K exhibit P_{trans} near the equilibrium value (0.48 ± 0.15), compared with only 14% of the dihedrals after complete (or nearly complete) β-relaxation $(5\tau_\beta)$. Figure 8.44 reveals that after τ_α all dihedrals contribute to the relaxation of the alkyl TACF for the LB-PBD melt at 140 K, and that 60% of the dihedrals are completely relaxed, compared to only 15% after $5\tau_\beta$ at the same temperature. A likely explanation for the partial relaxation associated with the β-relaxation process is conformational "memory" due to restricted chain motion imposed by the surrounding matrix. This conformational memory results in biased (non-equilibrium) population of conformational states for *individual* dihedrals on time scales smaller than the α-relaxation time. Hence, while all (or nearly all) dihedrals are able to visit each conformational state during the β-relaxation process, an increasing majority of them (with decreasing temperature) have a propensity to return to a preferred conformational state. This biasing of conformational populations on time scales shorter than the α-relaxation time becomes more severe with decreasing temperature due to more restrictive packing (higher density) and hence more heterogeneous environments, resulting in the observed reduction of the strength of the β-relaxation process with decreasing temperature. This mechanistic interpretation is supported by Figure 8.45, where the mean-square displacement (MSD) of monomers and the TACF are shown for the

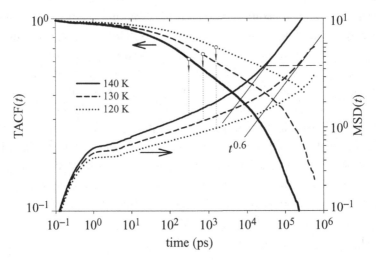

Figure 8.45 The alkyl dihedral TACF and monomer mean-square displacement (MSD) (Å^2) as a function of time for the three lowest temperature LB-PBD melts investigated. The horizontal dashed line denotes the length scale at which deviation from the expected scaling behavior for monomer displacements is observed, indicating the caging length scale. The vertical dotted lines indicate the β-relaxation time for each temperature. Data from Smith and Bedrov [54].

three lowest temperatures of the LB-PBD melt investigated as a function of time. Also shown are the β-relaxation times for the alkyl dihedrals and the expected scaling behavior for the monomer MSD in the diffusive regime. The onset of the diffusive regime is often considered to coincide with cooperative segmental relaxation (α-relaxation process). Deviation from diffusive motion indicates monomer caging by the matrix, which for the LB-PBD melts occurs on a length scale between $0.5\ \text{Å}^2$ and $5\ \text{Å}^2$ in the temperature range $120–140$ K, as indicated in Figure 8.45. On the time scale of the β-relaxation the monomer MSD is much less than $5\ \text{Å}^2$, indicating that the cage imposed by the matrix remains largely intact during the β-relaxation process. This cage, while not restricting access of individual dihedrals to each conformational state, biases the populations of individual dihedrals (but not the overall population when averaged over all dihedrals!). The final decay of the TACF (equilibration of individual dihedral populations) occurs after significant matrix motion results in the disappearance of the original caging-induced biasing of dihedral populations.

References

[1] A. Hoffmann, F. Kremer, and E. W. Fischer, *Physica A*, **201**, 106 (1993).
[2] S. P. Bravard and R. H. Boyd, *Macromolecules*, **36**, 741 (2003).

[3] A. Schoenhals, in *Dielectric Spectroscopy of Polymeric Materials*, edited by J. P. Runt and J. J. Fitzgerald (Washington, DC: American Chemical Society, 1997), Chapter 3. In turn adapted from F. Kremer, A. Hofmann, and E. Fischer, *Am. Chem. Soc. Polymer Preprints*, **33**, 96 (1992).

[4] S. Sen, Dielectric relaxation in aliphatic polyesters, Ph.D., Dissertation, University of Utah (2001).

[5] S. Havriliak and S. J. Havriliak, in *Dielectric Spectroscopy of Polymeric Materials*, edited by J. P. Runt and J. J. Fitzgerald (Washington, DC: American Chemical Society, 1997), Chapter 6.

[6] R. H. Boyd and C. H. Porter, *J. Polymer Sci.: Part A-2*, **10**, 647 (1972).

[7] C. H. Porter and R. H. Boyd, *Macromolecules*, **4**, 589 (1971).

[8] G. Williams, *Trans. Faraday Soc.*, **61**, 1564 (1965).

[9] S. Yano, R. R. Rahalkar, S. P. Hunter, C. H. Wang, and R. H. Boyd, *J. Polymer Sci., Polym. Phys. Ed.*, **14**, 1877 (1976).

[10] A. Arbe, U. Buchenau, L. Willner, D. Richter, B. Fargo, and J. Colmenero, *Phys. Rev. Lett.*, **76**, 1872 (1996).

[11] A. Arbe, D. Richter, J. Colmenero, and B. Farago, *Phys. Rev. E*, **54**, 3853 (1996).

[12] G. D. Smith, D. Bedrov, and W. Paul, *J. Chem. Phys.*, **121**, 4961 (2004).

[13] W. Paul, G. D. Smith and D. Bedrov, *Phys. Rev. E.*, **74**, 02150 (2006).

[14] K. Schmidt-Rohr and H. W. Spiess, *Multidimensional Solid-State NMR and Polymers* (London: Academic Press, 1996).

[15] U. Tracht, M. Wilhelm, A. Heuer, H. Feng, K. Schmidt-Rohr, and H. W. Spiess, *Phys. Rev. Lett.*, **81**, 2727 (1998).

[16] U. Pschorn, E. Roessler, H. Sillescu, S. Kaufmann, D. Schaefer, and H. W. Spiess, *Macromolecules*, **24**, 398 (1991).

[17] D. Schaeffer, H. W. Spiess, U. W. Suter, and W. W. Fleming, *Macromolecules*, **23**, 3431 (1990).

[18] A. Heuer, M. Wilhelm, H. Zimmermann, and H. W. Spiess, *Phys. Rev. Lett.*, **75**, 2851 (1995).

[19] M. D. Ediger, *Annu. Rev. Phys. Chem.*, **51**, 99 (2000).

[20] P. V. K. Pant, J. Han, G. D. Smith, and R. H. Boyd *J. Chem. Phys.*, **99**, 597 (1993).

[21] R. H. Boyd, R. H. Gee, J. Han, and Y. Jin, *J. Chem. Phys.*, **101**, 788 (1994).

[22] D. Rigby and R. J. Roe, *J. Chem. Phys.*, **87**, 7285 (1987).

[23] D. Rigby and R. J. Roe, *J. Chem. Phys.*, **89**, 5280 (1988).

[24] D. Rigby and R. J. Roe, *Macromolecules*, **22**, 2259 (1989).

[25] D. Rigby and R. J. Roe, *Macromolecules*, **23**, 5312 (1990).

[26] D. Rigby and R. J. Roe, in *Computer Simulation of Polymers*, edited by R. J. Roe (New York: Prentice Hall, 1991), Chapter 6.

[27] H. Takeuchi and R. J. Roe, *J. Chem. Phys.*, **94**, 7446 (1991).

[28] H. Takeuchi and R. J. Roe, *J. Chem. Phys.*, **94**, 7458 (1991).

[29] M. Muthukumar, *Adv. Chem. Phys.*, **128**, 1 (2004).

[30] I. Dukovski and M. Muthukumar, *J. Chem. Phys.*, **118**, 6648 (2003).

[31] M. Muthukumar, *Phil. Trans. Roy. Soc.*, **A361**, 539 (2003).

[32] G. Rutledge and S. Balijepalli, *J. Chem. Phys.*, **109**, 6523 (1998).

[33] G. Rutledge, M. S. Lavine and N. W. Waheed, *Polymer*, **44**, 1771 (2003).

[34] S. Toxvaerd, *J. Chem. Phys.*, **93**, 4290 (1990).

[35] P. Padilla and S. Toxvaerd, *J. Chem. Phys.*, **94**, 5650 (1991).

[36] J. Han R. H. Gee, and R. H. Boyd, *Macromolecules*, **27**, 7781 (1994).

[37] R. H. Boyd, *Trends Polymer Sci.*, **4**, 12 (1996).

[38] R. J. Roe, in *Atomistic Modeling of Physical Properties*, Advances in Polymers Science, Vol. 116, edited by L. Monnerie and U. Suter (New York: Springer Verlag, 1994), pp. 114–144.

[39] Y. Jin and R. H. Boyd, *J. Chem. Phys.*, **108**, 9912 (1998).

[40] E. Helfand, Z. R. Wasserman, and T. Weber, *Macromolecules*, **13**, 526 (1980).

[41] D. B. Adolf and M. D. Ediger, *Macromolecules*, **24**, 5834 (1991).

[42] M. D. Ediger and D. B. Adolf, *Macromolecules*, **25**, 1074 (1992).

[43] I. Zuniga, I. Bahar, R. Dodge, and W. L. Mattice, *J. Chem. Phys.*, **95**, 5348 (1991).

[44] E. Helfand, *J. Chem. Phys.*, **54**, 4651 (1971).

[45] W. Pechhold, S. Blasenbrey, and S. Woerner, *Colloid Polym. Sci.*, **189**, 14 (1963).

[46] R. H. Boyd and S. M. Breitling, *Macromolecules*, **7**, 855 (1974).

[47] R. H. Boyd, *J. Polymer Sci. Polymer Phys. Ed.*, **13**, 2345 (1975).

[48] R. H. Boyd and P. J. Phillips, *The Science of Polymer Molecules* (New York: Cambridge University Press, 1996).

[49] W. Jin and R. H. Boyd, *Polymer*, **43**, 503 (2002). The time scales in Figures 2–5 in this paper were mislabeled and should read from 0 to 30 ns in each.

[50] G. D. Smith, O. Borodin, and W. Paul, *J. Chem. Phys.*, **117**, 10350 (2002).

[51] R. H. Boyd and W. Jin, *ACS PMSE Abstracts*, August (2001).

[52] W. Paul and G. D. Smith, *Rep. Prog. Phys.*, **67**, 1117 (2004).

[53] A. Aouadi, M. J. Lebon, C. Dreyfus, *et al.*, *J. Phys.: Condens. Matter*, **9**, 3803 (1997).

[54] G. D. Smith and D. Bedrov, *J. Polym. Sci. Part B: Polym. Phys.*, **145**, 627 (2007).

[55] A. Döß, M. Paluch, H. Sillescu, and G. Hinze, *J. Chem. Phys.*, **117**, 6582 (2002).

[56] G. D. Smith and D. Bedrov, *J. Non. Cryst. Solids*, **352**, 293(2006).

Part III

Complex systems

There is a wide variety of polymeric systems in which relaxation processes play an important role but it would not be practical to attempt to cover them comprehensively in a meaningful way. Two types of systems, semi-crystalline polymers and miscible blends, have been chosen for exposition. The first type of system deals with *structural* complexity. Semi-crystalline polymers, even if chemically homogeneous, invariably are complex in consisting of two phases. Although the phases are distinct they interact with each other strongly and relaxational behavior reflects this. Molecular modeling has played a role in understanding these interactions. Miscible blends on the other hand are by definition structurally homogeneous. However, they are quite complex in that the constituent molecule types interact with each other and the results can be unexpected. Simulations can play an important role in probing these effects.

9

Semi-crystalline polymers

Long polymer chains consisting of identical monomeric units are, in principle, capable of being organized into crystalline arrays. Usually the chains are parallel bundles and the unit cell is based in some elementary way on the monomeric repeat unit. However, because of their great chain lengths it is kinetically difficult for polymers to form large crystals or to crystallize completely. In the case of quiescent crystallization from the melt the common morphology involves very thin lamellar crystals in which the crystallizing chains fold back and forth across the growth face. A given chain may be incorporated into several lamellae where the latter are organized in ribbon-like sheaves. It is inevitable that a considerable fraction of chain units will be constrained from being laid down on the growth faces. An appreciable fraction of uncrystallized material results. The local organization is that of stacked lamellae separated by amorphous layers (Figure 9.1). Thus crystallizable polymers are typically *semi-crystalline* two-phase systems. The degree of crystallinity, i.e., the volume of the crystal phase relative to the specimen volume can vary according to the crystallization conditions. Slow cooling versus rapid quenching, annealing etc. are typical variables. Higher degrees of crystallinity tend to be accommodated by thicker crystal lamellae with a relatively minor role for the amorphous interlayer thickness,

Questions of interest that arise are the following. To what extent does the amorphous fraction in a semi-crystalline environment resemble the bulk amorphous phase in respect to its relaxation behavior? Are there primary and secondary relaxations similar to the bulk amorphous polymer relaxations? Does the crystalline fraction have relaxation processes of its own and if so are they influenced by the amorphous fraction? How are experimentally observed relaxation processes to be assigned to one fraction or the other? Because the amorphous interlayer is quite thin, 5–10 nm or so, and connected to the crystals via chains entering and leaving both domains, it will not be surprising to find profound influences of the two phases upon each other.

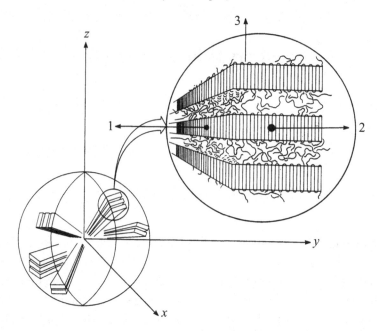

Figure 9.1 Spherulitic morphology. Bundles of stacked ribbon-like lamellae radiate from a primary nucleation site. Significant non-crystalline amorphous material separates the lamellae [1].

9.1 Phase assignment

9.1.1 Phase assignment – mechanical relaxation

Given the experimental observation of a relaxation process of unknown morphological origin, an obvious approach to its phase assignment would be to make measurements on specimens of varying crystallinity (as judged by density perhaps) and observe the effects on the relaxation signature. For example, a supposed amorphous fraction relaxation should show signs of weakening as the crystalline fraction increases. In Sec. 6.3 it was found that, because of the great difference between the relaxed and unrelaxed moduli, it was difficult to decide on a proper measure of the mechanical relaxation location for comparison with other methods such as dielectric relaxation. Similarly, in the present case the great difference between amorphous phase moduli and crystal moduli makes it difficult to decide which descriptor to plot against crystallinity in order to make a phase assignment. Just as in Figure 6.13, some illustrative calculations are useful.

In Figure 9.1, if a stress is considered to be locally applied in the vertical direction, labeled "3," the stress is uniform and strains in the alternate crystalline and amorphous layers are additive. This leads to the Reuss formula [2] that gives a lower bound to the modulus, E_L, of a mixture of isotropic phases (although the crystal phase is certainly not locally isotropic),

$$1/E_L = V_1/E_1 + V_2/E_2. \tag{9.1}$$

If the stress is applied in the 1, 2-plane, the strains are uniform (no debonding) and the stresses are additive, and the Voigt formula [2] results for the upper bound, E_U,

$$E_U = V_1 E_1 + V_2 E_2, \tag{9.2}$$

where subscript "1" refers to the amorphous phase and subscript "2" to the crystalline one and V_1, V_2 are the respective volume fractions of the phases. Because the phase moduli E_1, E_2 are so disparate the bounds are very far apart and not useful per se. However, they can be combined into a useful mixing equation, due to Tsai *et al.* [3], with the introduction of an additional parameter, ξ, as follows:

$$E = (1 + \xi E_U)/(\xi + 1/E_L). \tag{9.3}$$

For $\xi \to 0$ the lower bound Reuss formula is recovered and for $\xi \to \infty$ the upper bound Voigt formula results.

In a dynamic mechanical experiment it is tempting to use the variation of loss peak height with the degree of crystallinity as an indicator for the phase assignment of a relaxation process. Then the question arises as to whether to use loss modulus or compliance as the appropriate indicator. Some sample calculations with the above mixing relation, eq. (9.3), are revealing. The equation can be extended to the dynamic case by regarding the phase moduli in eq. (9.1) and eq. (9.2) as dynamic moduli $E_1^*(\omega)$ and $E_2^*(\omega)$. Suppose there is a relaxation in the amorphous component but none in the crystalline one. Then, at some fixed frequency, let the crystal phase modulus be $E_2' = 1.0$ but set $E_2'' = 0.0$. Further, let the crystal phase be stiffer than the amorphous one by setting $E_1' = 0.1 E_2'$. Place a relaxation in the amorphous phase by setting $E_1'' = 0.1 E_1'$. Then the specimen modulus, E, in eq. (9.3) can be rationalized into its components, E', E'', at various values of the degree of crystallinity, V_2. In addition, the specimen compliance components, J', J'', can be computed from eq. (1.59) and eq. (1.60). The results are shown in Figure 9.2.

Intuitively it might be expected that, since the loss process resides in the amorphous phase, the specimen loss components would decrease with increasing crystallinity. This is indeed the case for J''. However, for the modulus E'', for three of the cases the loss actually *increases* somewhat with crystallinity over most of the range of the latter. Only in the pure Voigt upper bound limit, eq. (9.2), is there a monotonic decrease in loss as crystallinity increases. It is noted in Figure 9.3 that tan δ, which is the same for both the compliance and modulus cases, eq. (1.61), does decrease with increasing crystallinity for all values of the ξ parameter. However, the dependence is weak for the approach to lower bound behavior.

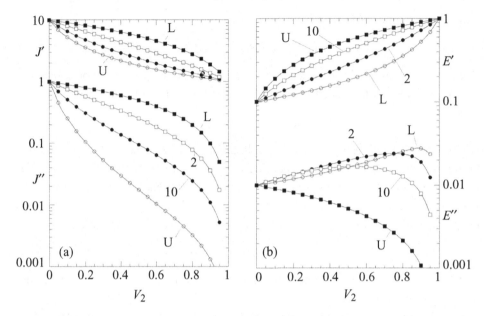

Figure 9.2   Specimen compliance J', J'' (a) and modulus E', E'' (b) plotted vs. degree of crystallinity, V_2 calculated at the points shown from the mixture equation, eq. (9.3). The relaxation is constrained to occur only in the amorphous phase by setting $E_1'' = 0.1E_1'$ and $E_2'' = 0.0$. The crystal phase is made stiffer than the amorphous phase by setting $E_1' = 0.1E_2'$. Four cases are shown: upper bound (labeled 'U') for $\zeta = \infty$, lower bound for $\zeta = 0$ (labeled L) and two intermediate cases $\zeta = 2, 10$.

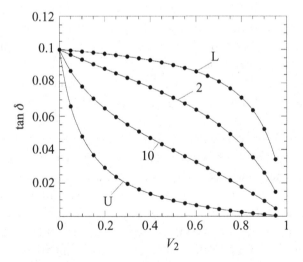

Figure 9.3   Specimen tan δ, vs. crystallinity for the same cases as in Figure 9.2. The ordinate scale is linear, however.

The loss modulus can indeed behave experimentally in a manner similar to that in Figure 9.2, i.e., the specimen loss modulus does not decrease with increasing crystallinity even though the loss process is clearly in the amorphous phase. Figure 9.4 shows torsion pendulum data for the shear modulus in the β subglass and α primary relaxation regions of PET specimens of varying crystallinity [4]. It can be seen that, especially in the β region, the loss modulus, G'', is quite insensitive to the degree of crystallinity. The data in the α region are somewhat compromised by the onset of melting, especially at the lower crystallinities.

The question arises then as to what could be a useful indicator of the morphological origin of relaxation. On examination of the β process in Figure 9.4 it may be seen that there are definite indicators in the real modulus, G', of phase origin. The difference in the low temperature unrelaxed modulus, G_u, and the high temperature relaxed modulus, G_r, in a given specimen appears to decrease noticeably as the crystallinity increases. In fact, phenomenologically, if there is a relaxation process in a given phase, it would be incumbent on the process that it possess relaxed and unrelaxed values of the relaxing property. This, in turn, suggests using the effect of crystallinity on relaxed and unrelaxed values, i.e., the relaxation strength as defined in Sec. 1.7, as an indicator that would be devoid of the problems with using the loss components. This is somewhat complicated by the fact the G_u, G_r quantities are themselves noticeably temperature sensitive. However, often it does turn out to be possible to effect a rather complete phenomenological analysis of an isochronal relaxation. In the case at hand it would proceed as follows. An empirical phenomenological model is adopted. In fact, the Cole–Cole function applied to dielectric relaxation, eq. (2.38), is useful when adapted to a relaxation function rather than a retardation function as

$$G^*(i\omega) = G_r + (G_u - G_r)(i\omega\tau_0)^\alpha / (1 + (i\omega\tau_0)^\alpha). \tag{9.4}$$

The relaxed and unrelaxed moduli, G_r, G_u, can be approximated as linear functions of temperature thus requiring two parameters for each, a value at a base temperature and a slope. The width parameter α also might be a linear temperature representation but a more complicated version with a slope increasing with temperature appears to be more effective. The relaxation time parameter, τ_0, is a two-parameter Arrhenius relation (eq. (6.6)) for a subglass process or a three-parameter VF, WLF type relation (eq. (6.7)) for a glass transition process. The parameters are determined by numerical regression of the calculated $G^*(i\omega)$ against the experimental data. Figure 9.5 shows the results of simultaneously fitting both the β subglass and α primary transition region in a semi-crystalline PET specimen [5]. This involves two additive versions of eq. (9.4), one for each process, that are linked through requiring the relaxed modulus of the lower temperature process, $G_r(\beta)$, to be equal to the unrelaxed value, $G_u(\alpha)$, for the higher temperature one, as in eq. (9.5), and

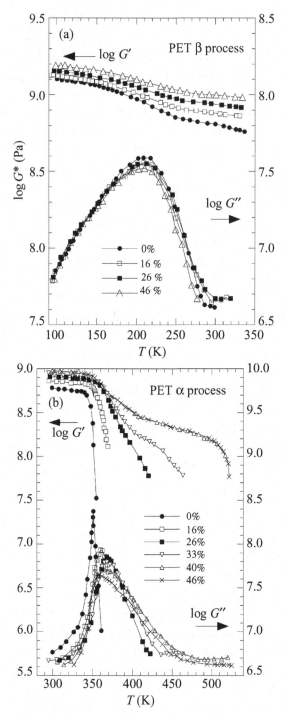

Figure 9.4 Complex shear modulus, G' (left-hand ordinates), G'' (right-hand ordinates) of PET in (a) the β- and (b) the α-relaxation regions at several crystallinities from 0% to nearly 50%. Torsion pendulum data, $\sim$1 Hz, of Illers and Breuer [4].

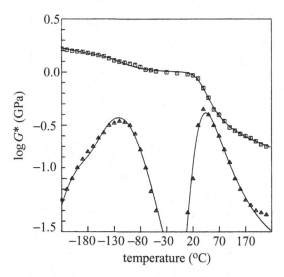

Figure 9.5 Phenomenological fitting of the dynamic shear modulus of a PET specimen of 46% crystallinity (see Figure 9.4). Torsion pendulum data ($\sim$1 Hz) of Illers and Breuer [4]. The log G'' (lower) curve and data are displaced 1.0 upward in plotting. The points are experimental and the curves are from eq. (9.5) with the parameters determined by numerical regression against the data points. Taken from Boyd [5] with permission.

where the τ_0, α parameters of course take on separate values for the α- and β-processes:

$$G^*(i\omega) = G_r(\alpha) + [G_r(\beta) - G_r(\alpha)] [(i\omega\tau_0)^\alpha / (1 + (i\omega\tau_0)^\alpha)]$$
$$+ [(G_u(\beta) - G_r(\beta))] [(i\omega\tau_0)^\alpha / (1 + (i\omega\tau_0)^\alpha)]. \qquad (9.5)$$

The relaxation strengths, $G_u - G_r$, determined from the fitting of both the α- and β-relaxations in several samples of varying crystallinity can then be employed as an indicator of phase origin. Because in PET, under normal crystallization conditions from the melt, the crystallinity range is restricted to $\sim$50% or less, the extrapolation to 100% crystallinity is a long one. Guidance from a mixing equation would be helpful in general in such extrapolations.

When developing better mechanical models for semi-crystalline polymers, advantage can be taken of the lamellar nature in formulating bounding equations that are much tighter than the Voigt and Ruess equations (eq. (9.1), eq. (9.2)) [6]. In general, it is found that lamellar lower bound equations give good fits to variations of limiting moduli, G_u and G_r, with crystallinity and in fact much better than the upper bound versions [7]. An illustrative calculation is shown in Figure 9.6.

Figure 9.7 shows the experimental limiting moduli, as determined by the phenomenological fitting process illustrated in Figure 9.5, for the PET specimens of

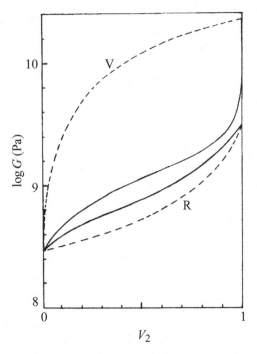

Figure 9.6 Calculated lamellar upper and lower bounds (solid curves) for the shear modulus of a macroscopically isotropic poly(ethylene) specimen vs. the degree of crystallinity. The crystal elastic constants are taken from the literature. The amorphous phase is assumed to be locally isotropic and a value for G_1 set at 0.3 GPa. The dashed lines are the Voigt upper bound (V) and the Ruess (R) lower bound. The upper and lower bounds diverge at 100% crystallinity due to the highly anisotropic nature of the crystal phase. Taken from Boyd [6] with permission.

varying crystallinity in Figure 9.4. The application of the lamellar lower bound equations to fitting the variation with crystallinity of the limiting moduli is also shown. In this procedure elastic constants for the crystal were estimated. For each curve the amorphous phase modulus is treated as a single adjustable parameter independent of crystallinity. It is apparent that the β subglass process is well accounted for as originating in the amorphous phase and disappears when extrapolated to 100% crystallinity. For the α primary transition region only the two highest crystallinities are useful due to the effects of the onset of melting, see Figure 9.4.

In summary, it is evident that the variation of the relaxation strength with crystallinity is the best indicator of phase origin. Thus it appears that, although considerable analysis is required, a reliable assignment of the phase origin in isochronal mechanical relaxation experiments on semi-crystalline polymers can be effected. The application to more complicated and contentious situations is described in Sec. 9.3.

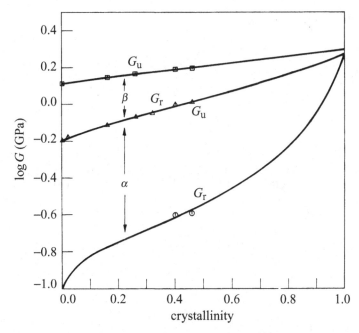

Figure 9.7 Variation with the degree of crystallinity of the limiting moduli G_r, G_u for both the α primary and β secondary processes in PET. The limiting moduli were determined by phenomenological fitting as illustrated in Figure 9.5. The curves are calculated from lamellar lower bound equations using fixed crystal elastic constants and an adjustable *crystallinity independent* amorphous phase modulus for each curve. Taken from Boyd [7] with permission.

9.1.2 Phase assignment – dielectric relaxation

Because, as described in Sec. 6.3, the difference between relaxed and unrelaxed dielectric constants is usually much less than is the case for mechanical moduli, the task of phase assignment is much easier than for the mechanical case. In addition, broadband frequency data are often available. Relaxation strength is still the best measure of phase origin. However, the phenomenological determination of relaxation strength via, for example, Argand plots (Sec. 2.5.3) is often straightforward. Figure 9.8 compares Argand plots for amorphous and semi-crystalline PET in the α primary transition region and Figure 9.9 shows the same comparison in the β secondary relaxation region.

Figure 9.10 displays the relaxation strengths determined from Argand diagrams for both the α- and β-processes in PET for specimens covering a wide range of crystallinities. It is apparent that the variation of relaxation strength is consistent with the amorphous phase origin for both processes. The α-process behavior of the lowest crystallinity, i.e., the essentially completely amorphous, sample, relative to the others is discussed in Sec. 9.2.1.

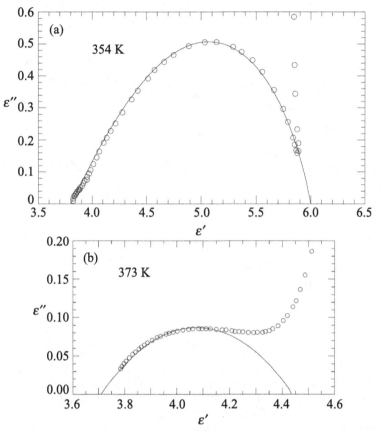

Figure 9.8 Comparison via Argand diagrams of the α-process relaxation strength in (a) amorphous PET and (b) semi-crystalline (50%) PET: in (a) the curve is an HN function fit; in (b) the curve is a Cole–Cole function fit. Data of Boyd and Liu [8].

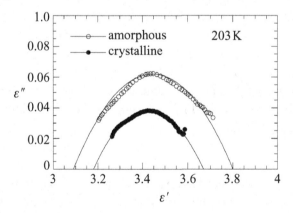

Figure 9.9 Comparison via Argand diagrams of the β-process relaxation strength in amorphous PET and semi-crystalline (50%) PET. Curves are Cole–Cole function fits. Data of Boyd and Liu [8].

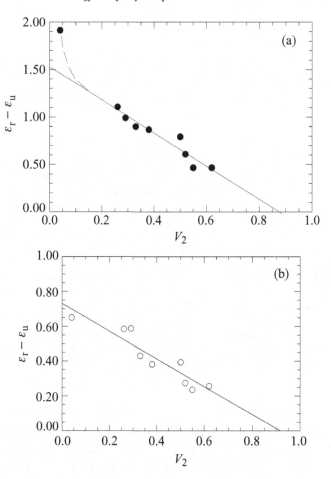

Figure 9.10 Relaxation strength variation with volume fraction crystallinity for: (a) the α primary relaxation, and (b) the β secondary relaxation, in PET. The lines are linear regressions. In (a) the lowest crystallinity specimen, which is essentially completely amorphous, is excluded from the regression. The dashed line serves to connect it with the other data (see text). Data of Coburn and Boyd [9].

9.2 Effect of crystal phase presence on amorphous fraction relaxation

9.2.1 Dielectric relaxation

The presence of the crystal phase has a dramatic effect on the α primary relaxation. Figure 9.11 compares the primary transition in semi-crystalline PET with that in amorphous PET. The ε'' loss spectrum in the crystalline specimen is broadened to the point of being difficult to resolve. In addition it is shifted several decades to lower frequency isothermally or to some 20 K higher temperature isochronally. Figure 9.12 displays these shifts in terms of an ε''_{max} loss map. However, the characteristic

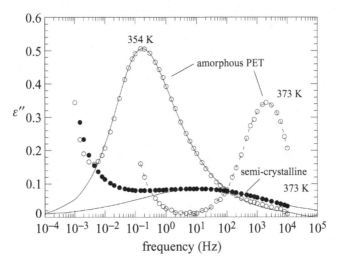

Figure 9.11 Effect of crystallization on the amorphous phase primary relaxation in PET. The filled points are the relaxation in the semi-crystalline (50%) environment at 373 K. The curve is a Cole–Cole function fit. The open points are the same process in amorphous PET. The solid curve (354 K) is an HN function fit. The dashed curve (373 K) is an interpolation function guide. The 363 K amorphous curve is at the same temperature as in the crystalline specimen, the 343 K amorphous curve is at a lower temperature. Data of Boyd and Liu [8].

VF, WLF curvature is retained. In addition to this effect on the dynamical behavior, there also appears to be a noticeable effect of the confinement by the crystals on the strength of the process. That is, reference to Figure 9.10 shows that in the α-process the wholly amorphous specimen does not lie on the regression line for the amorphous fraction in the semi-crystalline environment but noticeably above it. In other words the reduction in strength is somewhat greater than a simple proportionality to the amorphous phase volume fraction. This implies some immobilizing effect of the crystal phase on the amorphous phase that is static in nature and beyond the purely dynamic effect manifested in the broadening.

Turning to the β secondary relaxation, the situation as to the effect of crystal phase presence is quite different than that for the primary transition. Figure 9.13 compares the dielectric loss spectrum of a semi-crystalline PET with that of the wholly amorphous state. It is evident that the shape and location of the loss spectrum are very similar for the two states. Only the dilution effect of the crystalline phase on the relaxation strength is apparent. This is reinforced by the loss map in Figure 9.12 where, unlike for the primary relaxation, there is no significant difference between the semi-crystalline and the wholly amorphous materials with respect to loss locations and their temperature dependences. These effects are found to prevail over the complete crystallinity range available [9].

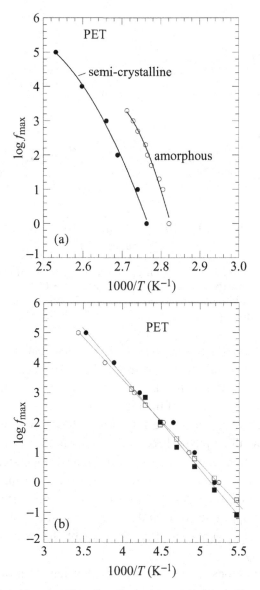

Figure 9.12 Loss maps showing the effect of the semi-crystalline environment on the time–temperature location of (a) the primary α-relaxation (data of Coburn and Boyd [9]) and (b) the secondary β-relaxation. In the latter, filled points are for a 50% crystalline specimen and open points are for an amorphous one (circles are data of Coburn and Boyd [9], squares are data of Boyd and Liu [8]).

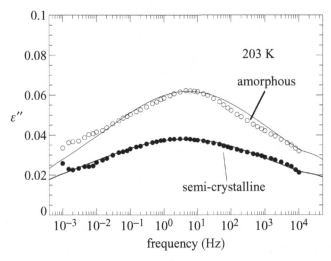

Figure 9.13 Dielectric loss spectrum of amorphous and semi-crystalline (50%) PET in the β secondary relaxation region. Curves are Cole–Cole function fits. The Cole–Cole α width parameters are 0.20 and 0.22 for the semi-crystalline and amorphous specimens respectively. Data of Boyd and Liu [8].

It is also of interest to note that in Figure 9.10 the relaxation strength for the β-relaxation for the essentially completely amorphous specimen does lie on the regression line for all the specimens, indicating no special effect of the crystal presence on the static properties of the relaxation.

In summary, the amorphous phase α primary relaxation is very sensitive to the presence of the crystalline phase, becoming much broader in the frequency plane and shifted to lower frequency isothermally or higher temperature isochronally. In contrast the subglass β secondary relaxation is, except for the dilution effect of occupying only part of the material, insensitive to the presence of the crystalline phase. These effects appear to be general in semi-crystalline polymers, although the completely amorphous polymer is often not available for comparison in polymers that crystallize easily.

The above observations are rationalized by consideration of the length scales involved in the two types of process. It was found in Chapter 8 that simulations show that, as temperature is lowered and the domain of restriction to inherently fast secondary relaxation is entered, relaxation becomes more and more dominated by localized conformational transitions that are able to take place with the involvement of only a few bonds. Therefore the length scales are quite small, of the order of 1 nm. Thus it seems quite reasonable that most of these transitions would occur without experiencing interference from the crystal phase. In contrast, at higher temperatures in the primary transition region where complete relaxation (i.e., complete exploration of conformational space) is achieved, the segmental motional scale must be considerably greater. Although it was argued in Sec. 6.3 that the effective length scale for dielectric relaxation in the primary transition region is much shorter than the molecular length, it would seem necessary that at least several persistence lengths (the distance required for a perpendicular dipole to lose correlation with dipoles in the same chain) would be required to allow complete relaxation in a given molecular segment. A typical persistence length is of the order of 1 nm or greater, so that several persistence lengths put the effective lengths in conflict with amorphous interlayer thicknesses of 5–10 nm. Partial restriction of a segment by crystal attachment or simply by proximity to the crystal surface would be expected to result in longer relaxation times or even exclude relaxation.

9.2.2 Mechanical relaxation

The presence of the crystal phase has an obvious effect on the composite semi-crystalline polymer. That is, a semi-crystalline specimen retains its mechanical integrity at temperatures above the primary transition in the amorphous phase and where a completely amorphous specimen would no longer be a solid. It is of interest, however, to inquire as to the mechanical properties of the amorphous interlayers and to what degree they depend on the degree of crystallinity. It has already been seen in Sec. 9.1.1 that crystallinity independent, amorphous, phase moduli could be invoked in fitting the limiting relaxed and unrelaxed moduli of the secondary and primary relaxations in PET (Figure 9.7). However, the mechanical integrity of the specimens restricts the relaxed primary transition modulus analysis to the higher crystallinities. But it is worth noting in Figure 9.7 that the invoked crystallinity independent, relaxed, amorphous, phase modulus, G_r, is 0.1 GPa. It is tempting to regard the amorphous interlayer in the semi-crystalline environment as effectively a cross-linked elastomer. But it is indeed a stiff elastomer as typical elastomers have moduli one to two orders of magnitude smaller. Thus it is concluded that the confinement by the crystal phase does have a profound effect on the stiffness of the amorphous fraction. The effect of the crystalline environment on the amorphous phase dynamics is similar to that found dielectrically in that the process is very broad in time or frequency. In the fitting process for PET presented in Figure 9.5 the Cole–Cole α width parameter in eq. (9.5) for the α-process in the amorphous phase takes on values of ~0.16 [5]. This indicates a very broad process. In fact it is comparable to or less than values found dielectrically for subglass relaxations and less than that for the dielectric primary α-process in semi-crystalline PET (e.g., the Cole–Cole α parameter for the 373 K loss curve in Figure 9.13 is 0.30 [8]).

9.3 Relaxations in semi-crystalline polymers with a crystal phase relaxation

Many semi-crystalline polymers exhibit a relaxation process that has its origin in the crystal phase. Usually this is found in mechanical relaxation and isochronally at temperatures higher than the primary and secondary amorphous phase relaxations [7]. The crystal phase motions are detectable via suitable solid-state NMR techniques, especially 2D exchange experiments [10,11]. Dielectrically, a crystal phase relaxation is an unusual occurrence because of different requirements for activity. In the dipolar case, molecular motions must take place that result in changes of dipole orientation (eq. (AII.14)). In a polymeric crystal, motions that do not involve the formation of defects, and are therefore relatively unlikely, must involve screw axis motion. That is, motion that involves rotation about and translation along the c-axis. This results in the chain returning to crystallographic register with no net energy penalty ($\Delta U = 0$ in eq. (AII.15)). In the dielectric case, since there is no change in dipole orientation, such a motion is not dielectrically active. Mechanically, the screw axis motion moves the chain translationally by the c spacing and gives rise to the possibility of relieving the crystal surface induced constraints on the amorphous fraction. This softening of the amorphous phase then is manifested as a mechanical relaxation process. In what follows, linear poly(ethylene) is selected for examination as the example of a polymer which has a prominent crystal phase relaxation. The experimental situation with respect to mechanical relaxation including the secondary and primary relaxations in the amorphous fraction and the crystal related relaxation is described first.

9.3.1 Mechanical relaxation in PE

The mechanical data described are those of Illers [12] which were obtained with a torsion pendulum; these data are displayed in Figure 9.14. The data cover an exceptionally wide range of crystallinities, from 47% to 95%. In discussing the data it is necessary to introduce a notational change. Due to the custom of denoting relaxations with Greek symbols in descending order of temperature isochronally, the crystal phase related relaxation is denoted as α, the primary relaxation in the amorphous phase as β and the secondary amorphous relaxation as γ. The data in Figure 9.14 encompass all three relaxations.

The phenomenological analysis described in Sec. 9.1.1 and Figure 9.5 and Figure 9.7 has been applied to these data with the result shown in Figure 9.15. Interlinked Cole–Cole functions similar to eq. (9.5) were fit to the data by numerical regression. Isothermal dynamic mechanical compliance data [13] and creep compliance data [14] as functions of frequency or time are also available for PET in the α-process region. The parameters found from the data fitting procedure in Figure 9.15 can be used to calculate isothermal dynamic compliance or creep compliance

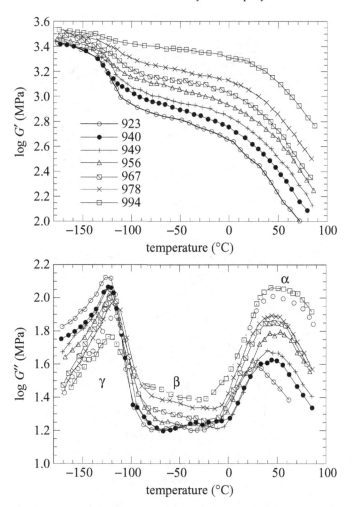

Figure 9.14 Dynamic shear modulus (~1 Hz) for poly(ethylene) over the temperature range that includes the α-, β-, and γ-relaxations and in samples that cover a wide range of specimen densities, shown in the key in kilograms per cubic meter, and hence crystallinities. Torsion pendulum data, ~1 Hz, Data of Illers [12].

for comparison with the experimental results. The result for the dynamic compliance is shown in Figure 9.16.

The relaxation strength parameters for each of the three relaxations occurring and parameterized in Figure 9.15 are shown as functions of the degree of crystallinity in Figure 9.17. It is very apparent in Figure 9.17 that all three relaxation processes behave as having an amorphous phase origin. Since the α-relaxation has been presented as having its origin in the crystal phase this obviously requires elucidation. In the introduction to this section the longitudinal mobility acquired by crystal phase

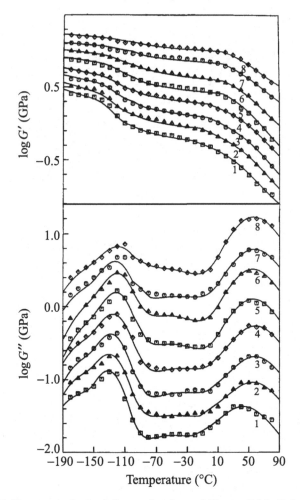

Figure 9.15 Phenomenological fits to the data of Figure 9.14. Points represent the data of Figure 9.14. For clarity each G' curve is displaced upward by 0.1 unit from the previous one (starting at curve 1); each G'' curve is displaced by 0.3. The curves are fits to the data using phenomenological functions as described in the text. Taken from Boyd [5] with permission.

chains and the companion effect on releasing amorphous phase constraints were alluded to. In order to describe this mobility in more detail it is necessary to inquire more closely into the elementary crystal phase chain motions. In this context the results of dielectric relaxation and NMR studies are particularly enlightening and these are taken up below.

9.3.2 Dielectric and NMR relaxation in poly(ethylene)

Although poly(ethylene) is inherently not polar, it can be rendered so without seriously altering its behavior by chemical decoration with a few dipoles; e.g., by

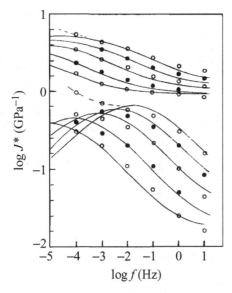

Figure 9.16 Dynamic compliance of poly(ethylene) in the α-relaxation region. Isothermal scans at 2.3, 15.5, 29.0, 40.4 and 50.5 °C. Representative points are from the data of Nakayasu *et al.* [13]. The curves are calculated from the same phenomenological parameters determined in the fitting in Figure 9.15. The dashed extensions show an experimental upturn due to an additional, higher temperature, lower frequency component in the α-process not observed in the torsion pendulum data (Figure 9.14). Taken from Boyd [5] with permission.

slight oxidation (carbonyl groups inserted in the chains) or chlorination (pendent Cl atoms) are examples. Figure 9.18 shows a comprehensive view of all three relaxation processes via isochronal scans for lightly oxidized poly(ethylene). Figure 9.19 displays a frequency scan of the loss behavior at 298 K, in which the α-process is in the center in the scan, and the companion Cole–Cole plot. A variety of experiments, such as variation of the degree of crystallinity, the melting, and the orientation of the specimens, show unequivocally that the dielectric α-process originates in the crystal phase [7]. The α-process is quite remarkable in that, for a polymer relaxation, it is extremely narrow. The Cole–Cole plot α width parameter in Figure 9.19 is 0.72, which is approaching SRT behavior.

As indicated, 2D exchange NMR experiments have been used to detect elementary crystal phase chain reorientations in a number of crystalline polymers [10]. However, in poly(ethylene), due to the planar zig-zag chain structure, the rotations during the flips are of necessity 180°. In this circumstance the relevant tensors are exactly inverted and flips are not detected. However, the rate of flips or barrier passages can be measured directly by dipolar ^{13}C NMR. In a ^{13}C–^{13}C bond pair the internuclear vector does reorient on chain rotation and can be detected [16]. Therefore a poly(ethylene) sample decorated dilutely via synthesis with such pairs

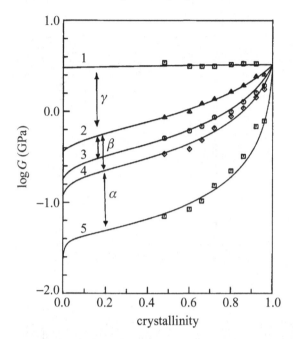

Figure 9.17 Relaxation strength as a function of the degree of crystallinity for the α-, β-, and γ-relaxations in poly(ethylene). Limiting moduli G_u, G_r for the γ process are given by curves 1, 2 respectively; for the β-process they are given by curves 2, 3, and 4, where 3 and 4 are the relaxed values at 175 and 300 K, respectively; curves 4 and 5 are for the α-process. Each curve is calculated using lamellar lower bound equations and an adjustable amorphous phase modulus that is the zero crystallinity ordinate. Taken from Boyd [5] with permission.

can measure the rate at which the containing chains flip, and with the advantage of no disturbance in the chemical structure of the chain. The correlation function from stimulated echo decays is found to be fit well by a KWW stretched exponential with a β parameter value of 0.8 [16]. Thus both the dielectric and NMR experiments indicate a very narrow process approaching single relaxation behavior.

The narrowness of the processes suggests that the chain reorientation involves a simple single barrier crossing associated with a rotational reorientation of the carbonyl or ^{13}C–^{13}C dipoles. As indicated, in poly(ethylene) this is accomplished by a 180° rotation of the entire polymer chain stem within the crystal lamella that also advances the chain by $c/2$, where c is the chain axis repeat distance. Since the dipoles are very dilute it is unlikely that there is more than one per chain stem and a net dipole rotation occurs. Thus dielectric and NMR active processes are observed. The remaining question concerns the activation energy associated with the chain stem rotation and hence the frequency–temperature location of the relaxation. The total energy barrier for a simultaneous concerted rigid rotation of the entire chain

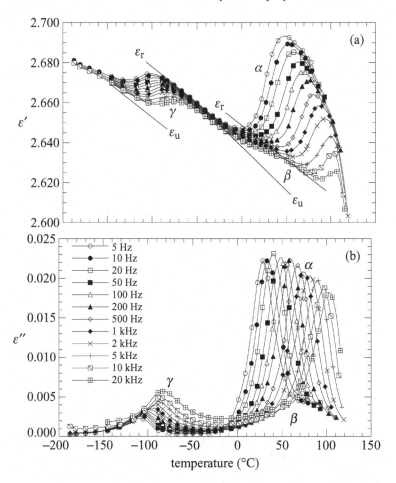

Figure 9.18 Dielectric permittivity (a) and loss (b) in lightly oxidized PE. Isochronal scans at the frequencies shown. Relaxed and unrelaxed dielectric constants drawn from high and low frequency envelopes are also shown for the β- and γ-processes. (The downward general slant in the permittivity envelopes is due to the effect of slight density change with increasing temperature on the electronic polarization permittivity of this very weak relaxation.) Data of Graff and Boyd [15].

stem is the sum of the substantial crystal mismatch energy from each CH_2 in the chain stem. It is too high to allow such a process in a thick crystal. In fact, as shown in Figure 9.20, the location of the α-relaxation does depend on crystal thickness, moving to lower frequency with increasing thickness. However, it does so in a manner that becomes less pronounced in thicker crystals.

The above crystal thickness dependence is rationalized by the fact that the chain need not rotate rigidly but the inherent torsional and extensional flexibility can allow only a part of the rotating chain to be seriously out of crystal register at a given

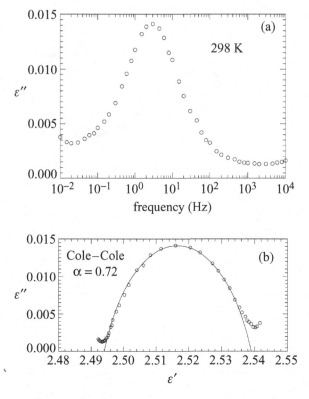

Figure 9.19 Dielectric loss in lightly oxidized poly(ethylene) for the α-process: (a) a frequency scan of the dielectric loss; (b) a Cole–Cole plot. The curve in (b) is a fit of the Cole–Cole function to the Cole–Cole plot. The α width parameter is shown. Data of Boyd and Liu [8].

instant of time during the overall rotation. Figure 9.21 illustrates a scheme that has been explored via detailed molecular mechanics calculations [17]. The twist in the chain initiates near the crystal surface and then proceeds to propagate across the crystal. The chain has sufficient longitudinal flexibility that the stems away from the moving twisted section maintain crystal register. The mismatch energy becomes mostly associated with the twisted portion. Migration of the twist to the opposite face completes the crystal chain stem dipolar reorientation and also the translation of the chain stem by $c/2$. The calculations show that once the twist is formed in a crystal that is much thicker than the twist extent it moves very freely. The activation barrier is essentially flat over the traverse through the crystal. Absolute rate theory can be used to predict the rate of passage over the barrier from the free energy of activation. The energetic component of the free energy is largely the barrier calculated from molecular mechanics and the entropic component can be estimated via a molecular mechanics calculation of the vibrational frequencies

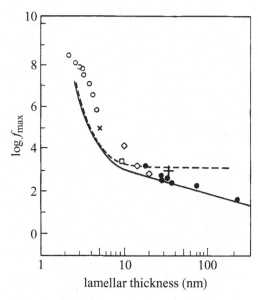

Figure 9.20 Dependence of crystal α-process location, as expressed by the frequency of maximum dielectric loss, in poly(ethylene) on lamellar thickness (at temperature 70 °C). Open circles are paraffinic ketones; cross is oxidized branched poly(ethylene); open square is single crystal oxidized poly(ethylene); open diamonds chlorinated poly(ethylene); filled circles oxidized poly(ethylene). The data references are given in Mansfield and Boyd [17]. The (+) point is from NMR in ^{13}C–^{13}C decorated HDPE [16]. The dashed curve is calculated on the basis of free streaming crossing of the crystal by a twist once formed. The solid curve incorporates a scattering or diffusive contribution in competition with free streaming that contains a single adjustable parameter. From Mansfield and Boyd [17].

of the twist compared to the straight chain. The rate of passage, k, i.e., the flip rate of the chains, is related to $\log f_{max}$ for a two-equal-sites model (eq. (AII.5)) as $2k = 1/\tau = \omega_{max} = 2\pi f_{max}$.

The dashed curve in Figure 9.20 is calculated based solely on the free streaming traverse feature. However, it is apparent that experimentally there continues to be some slowing of the process in very thick crystals. The solid curve incorporates a scattering or diffusive contribution in competition with free streaming that contains a single adjustable parameter [17]. Also included in Figure 9.20 is the NMR ^{13}C–^{13}C dipolar result at the same temperature. The agreement of the NMR result with the dielectric results is seen to be good. In addition the effect of temperature on the rate of flipping from the NMR method has been found to be in good agreement with that determined from the dielectric results [16,17]. The concordance of the two disparate methods is especially gratifying in the sense of giving credence to the supposed mechanism.

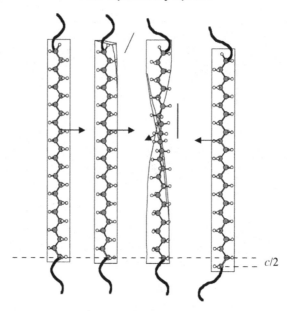

Figure 9.21 Crystal chain reorientation. Starting from left diagram, a twist initiates near the upper surface, the twisted region formed is fairly localized, it propagates to the lower surface leaving the chain in register but translated by half the repeat distance. A pendent dipole, indicated by arrows, undergoes 180° rotation. Taken from Boyd and Liu [8] with permission.

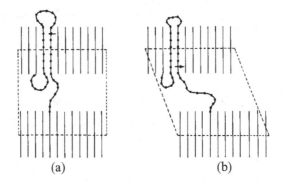

Figure 9.22 Schematic representation of the effect of mechanical strain taking place as the result of a taut tie chain (in (a)) lengthening by means of accumulated jumps in the crystal that shorten a loop on the opposite crystal face (in (b)). Taken from Boyd [19] with permission.

9.3.3 The crystal α-process and mechanical relaxation

It seems curious that, contrary to the crystal phase origin of the dielectric and NMR processes, for the mechanical α-process (Figure 9.17) is evidenced as an amorphous phase relaxation. The rationalization lies in the realization that elementary motions like those in Figure 9.21 do not lead to mechanical activity. According to

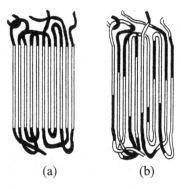

(a)   (b)

Figure 9.23 Schematic rendering of the chain diffusion resulting from accumulation of many elementary jumps as in Figure 9.21: (a) a starting state with ordered crystal chains (light) and disordered amorphous chains (dark); (b) the chains intermix as time proceeds. Taken from Schmidt-Rohr and Spiess [20] with permission.

Sec. AII.2, mechanical activity has to be associated with the stress biasing of the strain states before and after an elementary motion. However, motions such as those in Figure 9.21 do not result in a strain change within the crystal. The connections of the crystal surface to the amorphous phase lead to reorganization of the amorphous interlayer under applied stress [18], as depicted schematically in Figure 9.22. That is, the translational crystal motions are biased to increase the induced strain in the amorphous interlayer. This biasing, however, requires the statistical accumulation of many elementary crystal chain reorientations that can be described as diffusive longitudinal translation of the crystal chains (Figure 9.23).

In contrast in the dielectric and NMR cases one such jump reorients a dipole. Thus the mechanical and dielectric processes take place on somewhat different time or temperature scales. In Figure 9.14 the mechanical α peak ($\sim$1 Hz) occurs at $\sim$50 °C, and in Figure 9.19 the dielectric peak is seen to occur at $\sim$25 °C. Figure 9.24 shows, at the same temperature, 50 °C, the spectrum of relaxation times computed from Cole–Cole function fits to both the mechanical and dielectric data [19]. It is obvious that the mechanical process is much slower. In turn this is attributed to the necessity for statistical accumulation into longitudinal diffusion. Nevertheless the elementary jumps underlying the dielectric process are the same as those underlying the mechanical one and act as a necessary precursor. Thus, indeed, the mechanical α-process has its origins in the crystal phase even though the deformations take place in the amorphous phase.

9.4 NMR insights

Not only can NMR detect the crystal phase chain flips, as discussed above, but unlike the dielectric method, 2D exchange ^{13}C NMR directly detects the crystal

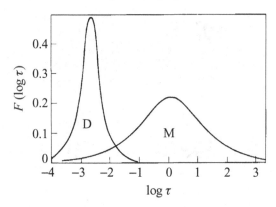

Figure 9.24 The spectrum of relaxation times at 50 °C for the dielectric (D) and mechanical (M) α-relaxation processes in lightly oxidized poly(ethylene). From Cole–Cole function fits of the respective processes. Taken from Boyd [19] with permission.

chain translational motions into and from the amorphous phase [20]. Key to the method is the fact that the resonances in the crystal and amorphous phases occur at different frequencies. In the 2D experiment (see Sec.3.4), the NMR frequencies before (ω_1) and after (ω_2) a waiting period are plotted in a plane with axes (ω_1, ω_2). The intensities of the signals are represented as heights in a 3D rendering or as contours in a 2D rendering. If there is no change in frequency, a point will lie on the diagonal. Changes in frequency occasioned by motion from one phase to the other during the waiting period will generate off-diagonal points and associated intensity. Figure 9.25 shows the result of an experiment on ultra high molecular weight poly(ethylene) where the excitation is provided by a series of single pulses accompanied by a recycle delay of 4 s. The very strong peaks are on the diagonal and are due to the separate crystal and amorphous phase resonances. Exchange between the phases during the waiting period shows up as the two off-diagonal bumps.

The 2D exchange experiments, because of the relatively short time periods necessary (a few seconds), are not effective in exploring the long-time nature of the crystal amorphous chain movements. However, analysis of 1D exchange experiments, where the delays can be varied over a broad range (1–2000 s), provides convincing evidence that the chain translational motions that result from accumulation of many rotational flips are diffusive in nature (as speculated above in contrasting the dielectric and mechanical processes) [20]. In this experiment the amorphous phase is selectively magnetized and the build-up of the intensity of the crystal resonance as a function of the recycle delay time is followed. For a classical free diffusion process the progress would follow the square root of time. In Figure 9.26 the intensity of the crystalline resonance is plotted against the square root of the recycle delay time for a number of temperatures. It is seen that pure square root dependence characteristic of free diffusion is not well obeyed and neither is a purely

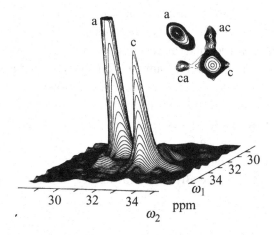

Figure 9.25 NMR 2D exchange experiment in poly(ethylene). In the 3D rendering, the axes ω_1, ω_2 represent the resonance frequencies before and after a waiting period determined by the pulse recycle delay (4 s). The upper right insert is a contour plot of the same information. The large peaks labeled a, c are on the diagonal and are resonances occurring in the amorphous and crystalline phases respectively. In the contour plot, the features ac and ca are resonances in the amorphous phase before the waiting period and in the crystal phase afterward and vice-versa (and are apparent in the 3D rendering as off-diagonal bumps). Taken from Schmidt-Rohr and Spiess [20] with permission.

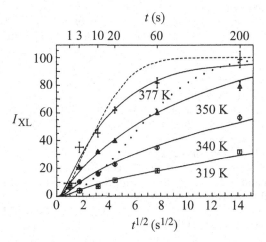

Figure 9.26 The intensity of the NMR crystal resonance in poly(ethylene) as a function of time after magnetization of the amorphous phase. The lower abscissa is the square root of the recycle delay time. The dashed curve represents a rate that matches the initial part of the 377 K results. The dotted curve represents a single exponential build-up. The solid curves are the results of a Monte Carlo simulation in which diffusion is partially inhibited by amorphous phase entanglements. Taken from Schmidt-Rohr and Spiess [20] with permission.

single exponential build-up. In order to incorporate an anticipated restriction on free diffusion by entanglements in the amorphous phase a Monte Carlo simulation was carried out that incorporated the restriction that loops cannot contract beyond the starting size by more than a fixed amount. This yields good agreement with the experimental results (Figure 9.26). In summary, the movement of the chains back and forth between the phases is well confirmed as a diffusive process and therefore its stress biasing provides the basis for the amorphous phase assignment of the mechanical α-relaxation process.

The chain motion associated with crystalline α-processes has proven to be a key factor in the drawability, especially ultra-drawability, of semi-crystalline polymers [21].

References

[1] J. C. Coburn, Ph.D. dissertation, University of Utah (1984).
[2] R. L. McCullough, in *Treatise on Materials Science and Technology*, Vol. 10 part B, edited by J. M. Schultz (New York: Academic Press, 1970).
[3] S. W. Tsai, J. C. Halpin, and N. J. Pagano, *Composite Materials Workshop* (Stamford, CT: Technomic., 1968).
[4] K. H. Illers and H. Breuer, *J. Colloid Sci.*, **18**, 1, (1963).
[5] R. H. Boyd, *Macromolecules*, **17**, 903 (1984).
[6] R. H. Boyd, *J. Polymer Sci. Polymer Phys. Ed.*, **21**, 493 (1983).
[7] R. H. Boyd, *Polymer*, **26**, 323 (1985). A comprehensive review of relaxations in semi-crystalline polymers with literature references.
[8] R. H. Boyd and F. Liu in *Dielectric Spectroscopy of Polymeric Materials*, edited by J. P. Runt and J. J. Fitzgerald (Washington, DC: American Chemical Society, 1997), Chapter 4.
[9] J. C. Coburn and R. H. Boyd, *Macromolecules*, **19**, 2238 (1986).
[10] K. Schmidt-Rohr and H. W. Spiess, *Multidimensional Solid-State NMR and Polymers* (New York: Academic Press, 1994).
[11] H. W. Beckham, K. Schmidt-Rohr, and H. W. Spiess, in *Multidimensional Spectroscopy of Polymers*, edited by M. W. Urban and T. Provder (Washington, DC: American Chemical Society, 1995), pp. 243–253.
[12] K. H. Illers, *Koll. Z. Z. Polym.*, **251**, 394 (1973).
[13] H. Nakayasu, H. Markovitz, and D. J. Plazcek, *Trans. Soc. Rheology*, **5**, 261 (1961).
[14] N. G. McCrum and E. L. Morris, *Proc. Roy. Soc. London, Ser. A.*, **292**, 506 (1966).
[15] M. S. Graff and R. H. Boyd, *Polymer*, **35**, 1797 (1994).
[16] W.-G. Hu, C. Boeffel, and K. Schmidt-Rohr, *Macromolecules*, **32**, 1611 (1999); also *Macromolecules*, **32**, 1714 (1999).
[17] M. Mansfield and R. H. Boyd, *J. Polymer Sci. Polymer Phys. Ed.*, **16**, 1227 (1978).
[18] N. G. McCrum, in *The Molecular Basis of Transitions and Relaxations*, edited by D. J. Meier (Midland Macromolecular Institute, 1978), p. 167.
[19] R. H. Boyd, *Polymer*, **26**, 1123 (1985).
[20] K. Schmidt-Rohr and H. W. Spiess, *Macromolecules*, **24**, 5288 (1991).
[21] W.-G. Hu and K. Schmidt-Rohr, *Acta Polymerica*, **50**, 271 (1999).

10

Miscible polymer blends

Polymer blends account for about 40% of all polymers produced [1]. Miscible polymer blends are technologically important because they have the potential to lead to new materials without the expense and time of new synthesis. Like the metal alloys that preceded them, polymer blends often benefit from synergistic effects that make them more useful than a simple linear combination of properties would anticipate. The importance of miscible polymer blends is likely to increase as a result of environmental and economic pressures. Growth areas for miscible polymer blends include recycled polymers where understanding of and the ability to improve the properties of mixtures of post-consumer polymers could greatly increase the utility of these materials. Furthermore, use of miscible polymer blends for drug delivery and biomedical applications, where blending of materials already approved for in vivo applications in order to achieve improved properties without the time and expense associated with obtaining approval for use of new polymer materials, is receiving increased attention. An improved understanding of miscible blends may also ultimately have significant impact on our understanding of other dynamically heterogeneous polymer systems such as polymer nanocomposites, microphase separated polymer blends, and polymers in confined environments.

The ability to understand, and ultimately to predict, the behavior of miscible polymer blends based upon the behavior of the pure component polymers that make up the blend is central to our ability to design polymer mixtures with desired properties. The dynamics and relaxation behavior of miscible polymer blends are particularly complicated and seem to defy a general predictive description. While calorimetry studies typically reveal a single, broad glass transition, probes of segmental dynamics reveal that each component in a thermodynamically miscible blend exhibits distinct dynamics. Furthermore, different experimental probes of segmental dynamics of miscible blends appear to give contradictory information about how blending influences the dynamics of the component polymers. Theoretical models attempting to explain the dynamical behavior of blends and their

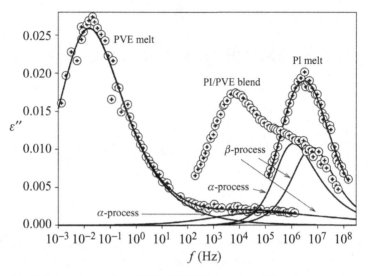

Figure 10.1 Dielectric loss in a 50/50 PI/PVE blend as well as the pure component polymer responses at 270 K. The solid lines are fits from Smith and Bedrov [3]. The symbols are experimental data from Arbe *et al.* [2].

components based on thermally driven concentration fluctuations, chain connectivity, or a combination of both have been proposed. In this chapter we discuss these models and their successes and short-comings in describing relaxation behavior in miscible polymer blends as well as insight into miscible polymer blend dynamics provided by MD simulations.

10.1 Poly(isoprene)/poly(vinyl ethylene) (PI/PVE) blends

NMR relaxation, dielectric spectroscopy, and QENS measurements of segmental dynamics in polymer blends comprising components with various degrees of thermodynamic compatibility, structural similarity, and dynamic asymmetry have led to the remarkable conclusion that component polymers in miscible blends, i.e., polymer mixtures that are intimately mixed from the viewpoint of static scattering, exhibit distinct segmental dynamics. Specifically, the component polymers of miscible blends exhibit segmental relaxation times that differ in magnitude and temperature dependence from each other and from their corresponding pure melt values. For example, Figure 10.1 shows the dielectric loss for a 50/50 blend of poly(isoprene) and poly(vinyl ethylene) (PI/PVE) at 270 K together with dielectric loss of pure PI and PVE melts at the same temperature [2,3]. The PI/PVE blend poses a severe challenge to our understanding of blend dynamics. 2D NMR measurements [4] for this blend show that upon blending the segmental dynamics in the low T_g component (PI) slow down relative to the pure PI melt while the segmental dynamics of the high T_g component (PVE) speed up by several orders of magnitude relative to its pure melt. In contrast, the dielectric response of the blend shown in Figure 10.1 is very broad and has a double peak structure with a low frequency peak positioned roughly between the peaks for the pure components while the second peak overlaps with the frequency range covered by the response of the pure PI melt. The latter can be interpreted (Sec. 10.2.2) to indicate that a significant fraction of the segmental relaxation for the PI component is similar to that observed in pure PI melt, i.e., it is uninfluenced by blending, which is inconsistent with 2D NMR results.

10.2 Models for miscible blend dynamics

Several models have been proposed to account for the influence of blending on the relaxation times and glass transition temperatures of the component polymers in a miscible blend as well as the change in the distribution of segmental relaxation times for the component polymers upon blending. These models assume that the influence of blending on segmental relaxation times and the glass transition temperature of the component polymers can be accounted for in terms of differing local concentrations of segments in some characteristic subvolume surrounding the segment of interest. The size of the subvolume and the physical mechanisms for determining the composition of the subvolume (and possible fluctuations in composition) depend on the model. In the Lodge–McLeish (LM) model [5] the length scale for the subvolume is on the order of the polymer chain Kuhn segment and the deviation of concentration (from blend average) within this subvolume due to chain connectivity accounts for the disparate segmental dynamics of the component polymers in a blend. In the concentration fluctuation (CF) models [6] the segmental dynamics of the chains are assumed to be determined by the composition of the "cooperative volume" surrounding the polymer segment. These cooperative volumes can either have a fixed size or be self-consistently determined by the local glass transition temperature which depends on the local CFs [7]. The latter model can, in principle, allow very broad distributions of local environments (concentration within the cooperative volume) that in turn can lead to a broad distribution of segmental relaxation times for blend components. Models that combine chain connectivity effects and concentration fluctuations within the relevant subvolume have been suggested [8].

10.2.1 The Lodge–McLeish (LM) model

The LM model [5] suggests that segmental dynamics are determined by the environment on the scale of the polymer statistical segment defined by the Kuhn length l_K.

On this length scale chain connectivity results in a local environment for a polymer segment that is richer in segments of its own type than the average blend composition. The effective local concentration around a segment can be defined as

$$\phi_{eff} = \phi_{self} + (1 - \phi_{self})\phi, \tag{10.1}$$

where ϕ is the blend composition (volume fraction of A or B segments) and ϕ_{self} is the so called "self-concentration" of the component under consideration (A or B) and corresponds to the volume fraction occupied by a polymer segment in the volume $V \sim l_K^3$. The model correlates ϕ_{eff} with a glass transition temperature $T_{g,eff}(\phi)$ for the local environment (quantified by ϕ_{eff}) of polymer segment, which is assumed to be equal to $T_g(\phi_{eff})$. To predict $T_g(\phi_{eff})$ the glass transition temperature of the pure components (T_g^A and T_g^B) can be used assuming that a mixing rule such as the Fox equation [9]

$$\frac{1}{T_{g,eff}(\phi)} = \frac{1}{T_g(\phi_{eff})} = \frac{\phi_{eff}^A}{T_g^A} + \frac{1 - \phi_{eff}^A}{T_g^B} \tag{10.2}$$

is applicable. Here ϕ_{eff}^A and ϕ_{eff}^B are the effective (local) concentrations (from eq. (10.1)) of A and B segments around the segment of interest. The LM model assumes that the segmental relaxations of the components in the blend are qualitatively (mechanistically) the same as in pure melts of the components and that only their temperature dependence shifts upon blending. If the segmental relaxation of a pure melt of the component A can be described by a VF temperature dependence,

$$\tau_{seg}^A(T) = \tau_\infty^A \exp\left[\frac{B^A}{T - T_0^A}\right], \tag{10.3}$$

then the temperature dependence of the segmental relaxation of this component in the blend can be represented by the same VF equation but with a concentration dependent T_0^A parameter determined as

$$T_0^A(\phi) = T_0^A + \left(T_g^A(\phi_{eff}) - T_g^A\right), \tag{10.4}$$

where T_g^A and $T_g^A(\phi_{eff})$ are the (known) glass transition temperature of the pure melt of polymer A and the (estimated using eq. (10.2)) glass transition temperature for the segmental relaxation of an A segment in the blend, respectively.

Figure 10.2(a) shows a comparison of the segmental dynamics of PS and PI chains as dilute components in a variety of glass-forming polymer hosts obtained by ^{2}H and ^{13}C NMR T_1 measurements [10]. The vertical axis of Figure 10.2(a) shows $\Delta T_{g,eff}$, the change in segmental dynamics for a PS or PI chain that is taken out of its pure melt and put into a host polymer matrix as a dilute chain while the horizontal axis shows the difference in T_g between the pure host and pure guest polymers. The dynamics of the dilute polymer chain is clearly correlated with the

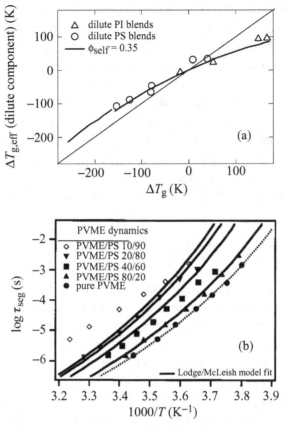

Figure 10.2 (a) The influence of blending on the segmental dynamics of dilute PS and PI in a variety of host polymers. The vertical axis shows the temperature shift between the dilute polymer dynamics in the blend and its pure homopolymer dynamics. The horizontal axis is the difference in T_g between the host and the pure guest (PS or PI melt). The bold line represents the segmental dynamics utilizing the LM model and the Fox equation (eqs. (10.1)–(10.4)) using $\phi_{self} = 0.35$. The thin line illustrates behavior expected when the dynamics of the guest polymer is slaved to that of the host. Data from Ediger *et al.* [10]. (b) The segmental relaxation time of PVME in PVME/PS blends from dielectric relaxation measurements. The lines are predictions of the LM model and the Fox equation (eqs. (10.1)–(10.4)) using $\phi_{self} = 0.82$. Reproduced from He *et al.* [11], with permission.

T_g of the host polymer. However, the dynamics of the dilute chain does not lie on the line with a slope of unity, indicating that it is not slaved to the host. Encouragingly, the LM model with $\phi_{self}^A = 0.35$ provides a reasonable description of the segmental dynamics of dilute PS and PI chains.

While Figure 10.2(a) shows that the LM provides a good description of the influence of blending on the segmental dynamics of dilute PI and PS chains in a variety of host polymers using a reasonable value of ϕ_{self}, the success of the LM model is far from universal. For example, segmental relaxation times for poly(vinyl methyl ether) (PVME) in PVME/PS blends obtained from dielectric relaxation measurements are shown in Figure 10.2(b) along with LM predictions [11]. When ϕ_{self} is adjusted to provide a good description of the segmental relaxation times of PVME in the blend with the lowest PS content, a poor description of the blends with higher content of the dynamically slow PS is obtained. Furthermore, the fit value of $\phi_{self} = 0.82$ is inconsistent with a value of $\phi_{self} \approx 0.25$ based upon the Kuhn segment length for PVME. The deviation between the LM prediction and actual PVME dynamics is quite severe for blends dilute in PVME, in contrast to Figure 10.2(a), where the LM model provides a good description of the influence of blending on segmental dynamics for blends dilute in PI and PS.

10.2.2 Concentration fluctuation (CF) models

CF models assume that segmental dynamics are determined by a local region that can be spontaneously rich in segments of either blend component due to thermal concentration fluctuations. The size of this local region is related to cooperatively rearranging volumes associated with the glass transition [6]. In the CF picture the segment of each component can have a broad distribution of local environments, and hence relaxation times, ranging from pure melt-like to dilute limits, depending on the thermodynamics of the mixture and the length scale of the cooperative region. The limitations of the CF concept can be clearly seen when the dielectric response for the PI/PVE blend (Figure 10.1) is analyzed. In one effort to explain the behavior of PI/PVE blends, the double peak shape in the dielectric loss of the PI/PVE blend was assumed to be due to two main relaxation processes operative in the blend [2]. The PI/PVE dielectric loss was represented as a superposition of Gaussian-like distributions of relaxation times representing the "fast" and "slow" relaxation processes with the peak in the distribution of the relaxation times for the slow relaxation process being between the segmental relaxation times observed for pure PVE and PI components and the peak for the distribution of relaxation times for the fast process being very close to the corresponding peak in the pure PI melt. In terms of the CF model, such an interpretation would imply that the dynamically relevant environment for PI in the PI/PVE blend must be such that the probability that this environment consists of pure PI segments is very high, which is clearly unphysical for any length scale in an intimately mixed 50/50 blend.

In a more recent attempt to describe the dielectric relaxation in PI/PVE blends, a simple lattice model was used to predict a distribution of local environments for the CF model [8]. In this work, concentration fluctuations were considered on a length scale comparable to the Kuhn length while experimental data on pure PI and PVE melts were used to obtain the parameters necessary for the model.

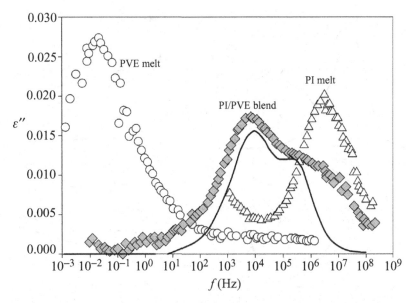

Figure 10.3 Predicted dielectric loss for a 50/50 PI/PVE blend (solid line) using the CF approach of Colby and Lipson [8] assigning a Debye relaxation process to each local environment. Experimental data are from Arbe *et al.* [2].

The prediction of the 50/50 PI/PVE blend dielectric loss obtained is shown in Figure 10.3. Unlike other CF models, where non-physical distributions of local environments have to be assumed to provide an adequate description of experimental data, the parameters and distribution functions in this work were physically reasonable. Unfortunately, the predictions of the CF model for the PI/PVE blend dielectric loss with these physical constraints do not capture the high frequency response of the blend.

10.3 MD simulations of model miscible blends

Insight into the underlying mechanisms of segmental relaxation in polymer blends that can be provided by MD simulations could greatly facilitate progress in developing models that correlate segmental dynamics with blend structure. A natural model system for such studies is a blend of CR-PBD and LB-PBD. The melt properties of these polymers have been discussed extensively in Chapters 6 and 8. LB-PBD has a lower T_g (108 K) than is observed for the CR-PBD (170 K) [12]. Hence, mixing of LB-PBD and CR-PBD chains provides a model polymer blend in which the components are quite different dynamically and yet have identical structure, thermodynamics, and conformations [13].

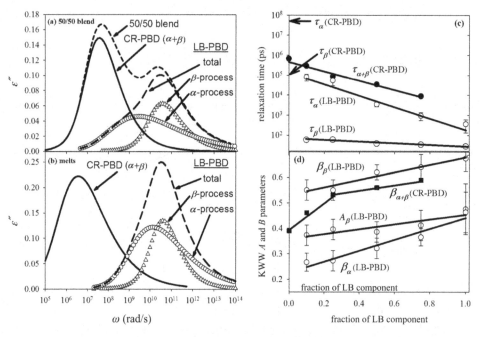

Figure 10.4 (a) Dielectric loss for a 50/50 CR-PBD/LB-PBD blend and the component polymers of the blend at 198 K. The CR-PBD component contribution is shown as a combined α- and β-relaxation, while the contributions of the individual α- and β-processes are shown for the LB-PBD component. (b) Dielectric loss for CR-PBD and LB-PBD melts at 198 K. Also shown are the contributions of the α- and β-relaxation processes to the pure melt dielectric response. (c) Integrated α- and β-relaxation times for the LB-PBD component and the relaxation time of the combined α- and β-process of the CR-PBD component of CR-PBD/LB-PBD blends as a function of blend composition. The solid lines are exponential fits that serve to guide the eye. Error bars reflect the sensitivity of the KWW fits of the DACF to the principal relaxation time parameter and in most cases are smaller than the symbol. (d) The amplitude of the β-relaxation process in the LB-PBD blend component (A_β) and the KWW stretching parameters for the combined α- and β-process in the CR-PBD component ($\beta_{\alpha+\beta}$) and the individual relaxation processes (β_α and β_β) in the LB-PBD component as a function of blend composition at 198 K. The solid lines are linear fits that serve to guide the eye. Error bars reflect the sensitivity of the KWW fits of the DACF to the respective parameter. Data from Bedrov and Smith [13].

In Figure 10.4(a) the dielectric loss of a 50/50 blend of CR-PBD and LB-PBD at 198 K is shown, while in Figure 10.4(b) the dielectric loss of the pure melts is shown. The dielectric loss is given as the Fourier transform of the DACF obtained from simulations [13]. The dielectric responses of the 50/50 CR-PBD/LB-PBD blend and the 50/50 PI/PVE blend (Figure 10.1) are remarkably similar. Also shown in Figure 10.4(a) are the contributions of the fast (LB-PBD) and slow (CR-PBD) blend components to the dielectric loss. The DACF for the LB-PBD component of the blend was fit as a sum of two KWW processes as was done for the LB-PBD melt (Sec. 8.2). While the DACF in CR-PBD melts can also be represented as a sum of two processes [13], as shown in Figure 10.4(b), it was found that the separation between the processes in CR-PBD when blended with LB-PBD was insufficient to allow for definitive assignment of strengths, relaxation times, and widths for the two processes. Hence, the combined α- and β-relaxation processes for the DACF in the CR-PBD component of the CR-PBD/LB-PBD blend are represented as a single KWW function.

Figure 10.4(a) reveals that the low-frequency peak in the dielectric response of the 50/50 CR-PBD/LB-PBD blend is due to the combined α- and β-relaxation processes in the CR-PBD component and the low frequency wing of the α-processes in the LB-PBD component, The α- and β-relaxation processes in the CR-PBD component are shifted to higher frequency while the β-relaxation process in the LB-PBD component is shifted to lower frequency and broadened upon blending. The high frequency peak in the blend response is due largely to the β-relaxation process of the LB-PBD component which is generally uninfluenced by blending combined with the high frequency wing of the broad α-relaxation process of the LB-PBD component.

10.3.1 Composition dependence of dielectric relaxation in CR-PBD/LB-PBD blends

Figure 10.4(c) shows the (integrated) relaxation times (eq. (6.13)) for the combined dielectric α- and β-relaxation processes in the CR-PBD component and the individual α- and β-relaxation processes in the LB-PBD component of a 50/50 CR-PBD/LB-PBD blend at 198 K as a function of blend composition. Also shown are the α- and β-relaxation times for the CR-PBD melt at 198 K where separation of the processes is sufficient to allow their resolution [13]. The insensitivity of the β-relaxation time for the LB-PBD component, the strong dependence of the α-relaxation time for the LB-PBD component, and the strong dependence of the relaxation time for the combined α- and β-relaxation processes for the CR-PBD component on blend composition can be clearly seen.

Figure 10.4(d) shows the amplitude (strength) of the dielectric β-relaxation process of the fast LB-PBD blend component, given as A_β in eq. (8.10), as well as the KWW stretching exponents for all resolvable processes in the blend as a function of blend composition. At 198 K (well above T_g of the pure LB-PBD melt) the strength of the β-relaxation process in the LB-PBD melt is comparable to the strength of the α-relaxation process. The strength of the β-relaxation process in the LB-PBD blend component changes very little with blending, indicating that the β-relaxation process of the fast component, while becoming well separated from the α-relaxation process in the blend (Figure 10.4(c)), remains strong for all blend compositions and is responsible for a significant fraction of the relaxation of the fast LB-PBD component in the blend. It can also be seen that both the α- and β-relaxation processes in LB-PBD broaden (i.e., the β stretching exponent becomes smaller) upon blending with the slow CR-PBD component. The broadening of the α-relaxation process in LB-PBD with blending is much more pronounced than that of the β-relaxation process, with the KWW β parameter for the α-process decreasing by almost 45% in the 10 wt% LB-PBD blend relative to the pure melt, compared with a decrease of only 20% in the KWW β parameter for the β-process.

For the CR-PBD component Figure 10.4(d) reveals significant narrowing of the combined α- and β-relaxation processes upon blending with the fast LB-PBD component, particularly for low LB-PBD content blends. This narrowing is likely due to a combination of reduced separation between the α- and β-relaxation processes with blending (note that in the pure melt at 198 K the processes in CR-PBD are separated by almost three orders of magnitude as shown in Figure 10.4(b)) and intrinsic narrowing of the individual processes [13]. The decrease in the width of the combined α- and β-relaxation in the CR-PBD component upon blending with the fast LB-PBD component for low LB-PBD blends may reflect a greater influence (speeding up) due to blending on the α-relaxation compared to the β-relaxation, resulting in a decrease in the separation of the processes. Further increase in LB-PBD content has relatively little influence on the width of the combined relaxation and hence on the separation and intrinsic width of the α- and β-relaxation processes in CR-PBD.

In summary, while blending with slow CR-PBD *increases* the separation between the α- and β-relaxation times for the fast LB-PBD component and results in broadening of the relaxation processes (particularly the α-relaxation process), blending of the slow CR-PBD component with the fast LB-PBD narrows the combined α- and β-relaxation, consistent with a *decrease* in the separation between the α- and β-relaxation times and an intrinsic narrowing of the relaxation processes. Both of these results indicate that blending has a stronger influence on the cooperative α-relaxation processes of the component polymers than on the local (main chain) β-relaxation processes.

10.3.2 *Temperature dependence of dielectric relaxation in a CR-PBD/LB-PBD blend*

Figure 10.5(a) shows the temperature dependence of the (integrated) dielectric β-relaxation times for the LB-PBD component of a CR-PBD/LB-PBD blend with

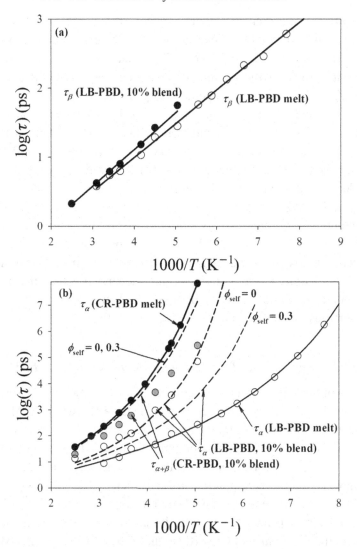

Figure 10.5 Integrated β-relaxation times (a) and α-relaxation times (b) for CR-PBD and LB-PBD melts and components in a CR-PBD/LB-PBD blend (10% LB-PBD) as a function of inverse temperature from KWW fits to the DACF. Lines in (a) are Arrhenius fits to the relaxation times. Solid lines in (b) are VF fits, while the dashed lines are predictions of the LM model combined with the Fox equation (eqs. (10.1)–(10.4)) for the indicated self-concentrations. Data from Bedrov and Smith [13].

10% LB-PBD. The β-relaxation times are very similar to those in LB-PBD melt, as illustrated in the figure, due to the insensitivity of the β-relaxation process to blending with the slow CR-PBD component. The β-relaxation times for LB-PBD can be well described by an Arrhenius temperature dependence in both the pure melt and the 10% LB-PBD blend. The temperature dependence of the dielectric α-relaxation time for the LB-PBD component in the 10% LB-PBD blend as well as those of the α-relaxation times for the pure CR-PBD and LB-PBD melts and the relaxation times for the combined α- and β-relaxation processes in the CR-PBD component of the blend are shown in Figure 10.5(b). The α-relaxation time for LB-PBD in the blend exhibits significantly stronger temperature dependence than is seen for the pure LB-PBD melt over the range of temperatures studied, similar to that of the combined α- and β-relaxation process of the majority blend component (the CR-PBD component). However, despite the dominance of the slow component (90 wt%) in the blend, the α-relaxation process for the LB-PBD components exhibits a *different* temperature dependence than the dominant CR-PBD component, consistent with experimental measurements of segmental dynamics in numerous miscible blends (e.g., see Figure 10.2).

10.3.3 Application of the LM model to a CR-PBD/LB-PBD blend

The ability of the self-concentration concept of the LM model to describe the influence of blending on segmental relaxation in the LB-PBD/CR-PBD blends has been investigated [13]. In Figure 10.5(b) predictions of the LM model as formulated in eqs. (10.1)–(10.4) for the α-relaxation times of the CR-PBD and LB-PBD components with $\phi_{self} = 0.3$, corresponding to the self-concentration yielded by a dynamically relevant volume corresponding to the Kuhn length in PBD, are shown. For the CR-PBD component the segmental relaxation times predicted by the LM model and their temperature dependence are quite similar to those seen for the pure CR-PBD melt due to the dominance of CR-PBD in the blend (90 wt%). The LM predictions for the segmental relaxation times for the CR-PBD component deviate significantly from the relaxation times for the combined α- and β-relaxation process observed in the blend. This deviation increases with decreasing temperatures. The separation between the α- and β-relaxation processes increases with decreasing temperature and it becomes more difficult to adequately/completely sample the contribution of the α-process to relaxation due to increasing α-relaxation times. As a consequence, the relaxation time obtained for the combined α- and β-relaxation process deviates more from the true segmental (α-)relaxation time with decreasing temperature. It is possible that the difference between LM predictions for the segmental relaxation time of blend components and observed relaxation times (experimental or simulation) in some blends may be due, at least in part, to the inability of the measuring technique to separate strong local β-relaxation contributions to the relaxation from the true segmental relaxation process (the α-process) and/or a measurement window that does not allow sufficient sampling of the long time contribution of the α-process to segmental relaxation.

For the LB-PBD blend component, where it is possible to separate relaxation due to the local and segmental processes and to completely sample the α-relaxation, it is clear that the LM model with $\phi_{\text{self}} = 0.3$ significantly underestimates the slowing down of segmental relaxation upon blending. Hence, for the LB-PBD component of the 90% CR-PBD/10% LB-PBD blend, the LM model with a physically reasonable ϕ_{self} underestimates the coupling between the segmental dynamics of the component and the dynamics of the matrix. $\phi_{\text{self}} = 0.0$ for the LB-PBD component does provide a good description of the α-relaxation times for the fast blend component, as illustrated in Figure 10.5(b), indicating, within the LM formulation, that the segmental relaxation of the fast component is completely slaved to that of the matrix, i.e., the local environment experienced by a segment is determined entirely by the bulk composition of the blend. However, application of the LM model with $\phi_{\text{self}} = 0.0$ is contrary to the spirit of the model which is predicated on the assumption that the local, dynamically relevant environment for a segment differs in composition from the bulk blend due to chain connectivity effects.

10.4 PI/PVE blends revisited

The above analysis of segmental relaxation processes in LB-PBD/CR-PBD blends leads to several important observations that may be applicable to relaxation mechanisms and segmental dynamics of a wide range of miscible polymer blends. Specifically, blending with a significantly slower component promotes separation between the α- and (main chain) β-relaxation processes of the fast component. The two processes may be nearly merged in the pure melt and hence exhibit, for example, a single, broad loss peak in dielectric studies. Separation is promoted by blending because the (local) β-relaxation process of the fast component is at most weakly affected by blending while the (cooperative) α-relaxation time of the fast component exhibits strong concentration dependence. The β-relaxation process in the fast component preserves its strength upon blending and can be comparable in strength to the α-relaxation process at temperatures well above the T_{g} of the fast component pure melt. The α-process of the fast blend component shifts to lower frequencies upon blending and can overlap with the α- and β-relaxation processes of the slow component.

It is instructive to revisit PI/PVE blends in light of these observations. As with the CR-PBD and LB-PBD melts, the dielectric response of pure PI and PVE melts can be represented as a sum of α- and β-relaxation processes [3]. Here, the experimental

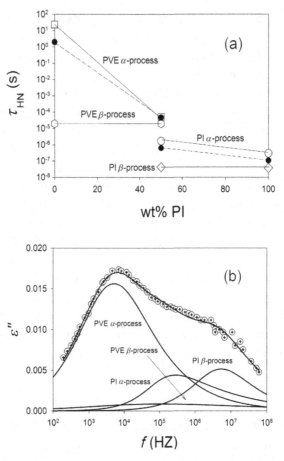

Figure 10.6 (a) Relaxation times for the dielectric α- and β-relaxation processes in pure PVE and PI melts as well as the 50/50 PVE/PI blend at 270 K. Dielectric relaxation times (open symbols) are principal relaxation times from HN fits to experimental blend data and are taken from Smith and Bedrov [3]. 2D NMR relaxation times (closed symbols) are extrapolated from experimental data of Chung *et al.* [4] obtained at lower temperatures. The lines serve to guide the eye. (b) Representation of the dielectric loss in a 50/50 PI/PVE blend as a sum of α- and β-relaxations in the component polymers (also shown). Data from Smith and Bedrov [3]. Experimental data for the blend (symbols) are from Arbe *et al.* [2].

dielectric losses for PI and PVE at 270 K have each been represented in terms of a sum of two HN functions

$$\varepsilon^*(f) = \Delta\varepsilon_\alpha \frac{1}{[1 + (2\pi i f \tau_a)^{\alpha_\alpha}]^{\gamma_\alpha}} + \Delta\varepsilon_\beta \frac{1}{[1 + (2\pi i f \tau_\beta)^{\alpha_\beta}]^{\gamma_\beta}} \qquad (10.5)$$

as shown in Figure 10.1. The corresponding relaxation times (τ_α and τ_β) for the pure melts are shown in Figure 10.6(a). The dielectric response of the 50/50 PI/PVE blend at 270 K is represented as

$$\varepsilon^*(f) = \left[\Delta\varepsilon_\alpha \frac{1}{[1 + (2\pi i f \tau_a)^{\alpha_\alpha}]^{\gamma_\alpha}} + \Delta\varepsilon_\beta \frac{1}{[1 + (2\pi i f \tau_\beta)^{\alpha_\beta}]^{\gamma_\beta}} \right]_{PVE}$$

$$+ \left[\Delta\varepsilon_\alpha \frac{1}{[1 + (2\pi i f \tau_a)^{\alpha_\alpha}]^{\gamma_\alpha}} + \Delta\varepsilon_\beta \frac{1}{[1 + (2\pi i f \tau_\beta)^{\alpha_\beta}]^{\gamma_\beta}} \right]_{PI}, \quad (10.6)$$

as shown in Figure 10.6(b). Also shown are the contributions of the α- and β-relaxation processes of the two components to the dielectric response of the 50/50 PI/PVE blend. The relaxation times for the α- and β-relaxation processes in the 50/50 PI/PVE blend are shown in Figure 10.6(a).

In representing the dielectric response of the 50/50 PI/PVE blend via eq. (10.6), it was assumed that the β-relaxation processes in the PI and PVE components are largely uninfluenced by blending. For PVE, the same strength, shape, and principal relaxation time as obtained for the PVE melt (Figure 10.1) were utilized for the blend. For PI, the strength of the β-relaxation process was determined simultaneously in the blend and the melt (along with that of the α-process) and was assumed to be uninfluenced by blending as was the shape of the β-relaxation. The principal relaxation time for the β-relaxation process in PI was an adjustable parameter in the fitting of the dielectric loss in the blend. The shape of the α-relaxation process in both components was allowed to change on blending as were the principal relaxation times. Extrapolation of segmental relaxation times obtained from 2D NMR measurements [4] at lower temperature to 270 K as well as diffusion measurements [14] at 270 K on 50/50 PI/PVE blends and PI melts indicate that the segmental dynamics and self-diffusion of PI slow down by a factor of about 6 upon blending. A constraint of $\tau_\alpha(PI)_{blend} = 6\tau_\alpha(PI)_{melt} = 1.71 \times 10^{-6}$ s was therefore introduced in the representation of the dielectric loss of the 50/50 PI/PVE blend via eq. (10.5). Here, $\tau_\alpha(PI)_{melt}$ is the principal relaxation time obtained from fitting of the pure melt dielectric loss (Figure 10.6(a)). No constraint was placed on $\tau_\alpha(PVE)_{blend}$. The relaxation strength for the α-relaxation process for PVE in the blend was taken as that obtained for the pure melt. For PI, the strength of the α-relaxation process was determined simultaneously in the blend and the melt (along with that of the β-process) and was assumed to be uninfluenced by blending.

Dielectric relaxation in the PVE and PI components of the 50/50 PI/PVE blend as represented in Figure 10.6(b) is consistent with the behavior of CR-PBD and LB-PBD observed in simulations of the 50/50 CR-PBD/LB-PBD blend enumerated above as well as in 2D NMR studies of segmental dynamics in a 50/50 PI/PVE blend. Specifically, the relaxation time for the α-relaxation process of the PVE component is strongly influence by blending. After blending, a difference of a factor of 30 between the α-relaxation times for the PVE and PI components is observed, consistent with 2D NMR measurements where extrapolation from lower temperatures yields a difference of around a factor of 70 at 270 K (Figure 10.6(a)). The correspondence of segmental relaxation times from 2D NMR studies and the α-relaxation times from dielectric measurements for the PI component appears to indicate that for the PI component 2D NMR measurements are sensitive primarily to the α-relaxation of the component polymers, consistent with the inability of this technique to access relaxations on time scales faster than a microsecond. Additionally, the PVE α-relaxation process is observed to narrow moderately and the PI α-relaxation process to broaden upon blending, consistent with observations for the 50/50 CR-PBD/LB-PBD blend. Finally, the β-relaxation processes are largely uninfluenced by blending. While this was assumed in fitting eq. (10.6) to the experimental loss data for the blend, the accurate representation of the dielectric loss of the blend (Figure 10.6(b)) supports this assumption. Furthermore, the insensitivity of local relaxation processes to blending has been observed for glass forming molecular mixtures [15,16] and in some miscible polymer blends [17].

The representation of dielectric loss in the PI/PVE blend shown in Figure 10.6(b) reveals that the high frequency dielectric loss in the PI/PVE blend is due largely to the β-relaxation of the dynamically fast PI component that is strong (large amplitude) at temperatures above the glass transition of pure PI and whose shape, strength, and principal relaxation time are largely uninfluenced by blending, as was found for the LB-PBD/CR-PBD blend. There is also a contribution to the high frequency loss of the blend from the α-relaxation process in the PI component which has broadened somewhat compared to the pure PI melt with blending. Examination of Figure 10.4(d) indicates that, for the LB-PBD/CR-PBD blend, the α-relaxation in the fast (LB-PBD) component continues to broaden with increasing content of the slow (CR-PBD) component for the entire range of compositions investigated. Hence, the broadening of the α-relaxation process in the fast component is not consistent with the concentration fluctuation model for miscible blends since we would anticipate the broadest distribution of local environments, and hence broadest relaxation, to occur for the 50/50 blend. Furthermore, as with the combined α- and β-relaxation process in the CR-PBD component of the LB-PBD/CR-PBD blends, the α-relaxation of the slow (PVE) component of the PVE/PI blends appears to *narrow* upon blending with the fast (PI) component, counter to the predictions of the CF model. Hence, while the cause of the broadening of the α-relaxation of the fast blend component (LB-PBD, PI) upon blending with the slow component (CR-PBD, PVE) is unknown, it does not appear to be due to local concentration fluctuations. One possible source of this broadening is a changing relationship between the cooperative α-relaxation, which is strongly influenced by the (ever slowing) environment upon blending with the slow component, and the β-relaxation, which is not strongly influenced by blending and remains very strong. This behavior is in stark contrast to that observed in the pure PI [3] and LB-PBD (Chapter 8) melts with decreasing temperature, where the slowing matrix dynamics that accompany decreasing temperature (as opposed to those associated with blending with a slow polymer) do not foment broadening of the α-relaxation but are associated with a reduction in the strength of the β-relaxation process.

Unlike previous attempts to explain the behavior of PI/PVE blends, the representation given in Figure 10.6(b) does not invoke the concept of CFs to account for the dynamically heterogeneous response of the blend. The representation of the dielectric loss in the 50/50 PI/PVE blend as a sum of contributions from the α- and β-relaxation processes of the component polymers is better than can be obtained using the CF approach with a reasonable distribution of local environments, such as shown in Figure 10.3, while an improved representation of the high frequency loss in the context of concentration fluctuations requires an unphysically high probability of pure PI domains [7]. The significant high frequency loss of the PI/PVE blend appears to be due largely to the β-relaxation of the PI component that does not shift with blending, as opposed to a significant fraction of the PI α-relaxation that is not influenced by blending due to a PI-rich environment. In summary, the response of the 50/50 PI/PVE blend at 270 K seems to be due to the very different response of the cooperative α-relaxation and local β-relaxation to blending and not to the dynamically heterogeneous response induced by local compositional fluctuations.

References

[1] L. A. Utracki (ed.), *Polymer Blends Handbook* (Boston: Kluwer Academic, 2002).

[2] A. Arbe, A. Alegria, J. Colmenero, S. Hoffmann, L. Willner, and D. Richter, *Macromolecules*, **32**, 7572 (1999).

[3] G. D. Smith and D. Bedrov, *Eur. Polym. J.*, **45**, 191 (2006).

[4] G.-C. Chung, J. A. Kornfield, and S. D. Smith, *Macromolecules*, **27**, 5729 (1994).

[5] T. P. Lodge and T. C. B. McLeish, *Macromolecules*, **33**, 5278 (2000).

[6] A. Zetsche and E. W. Fischer, *Acta. Polym.*, **45**, 168 (1994).

[7] S. K. Kumar, R. H. Colby, S. H. Anastasiadis, and G. Fytas, *J. Chem. Phys.*, **105**, 3777 (1996).

[8] R. H. Colby and J. E. Lipson, *Macromolecules*, **38**, 4919 (2005).

[9] T. G. Fox, *Bull. Am. Phys. Soc.*, **1**, 123 (1956).

[10] M. D. Ediger, T. R. Lutz, and Y. He, *J. Non-Crystal. Solids*, in press.

[11] Y. He, T. R. Lutz, and M. D. Ediger, *J. Chem. Phys.*, **119**, 9956 (2003).

[12] G. D. Smith and D. Bedrov, *J. Non-Crystal. Solids*, **352**, 4690 (2006).

[13] D. Bedrov and G. D. Smith, *Macromolecules*, **39**, 8526 (2006).

[14] J. C. Haley, T. P. Lodge, Y. He, M. D. Ediger, E. D. von Meerwall, and J. Mijovic, *Macromolecules*, **36**, 6142 (2003).

[15] K. Ngai and S. Capaccioli, *J. Phys. Chem. B*, **108**, 11118 (2004).

[16] T. Psurek, S. Maslanka, M. Paluch, R. Nozaki, and K. L. Ngai, *Phys. Rev. E*, **70**, 011503 (2004).

[17] I. Cendoya, A. Alegria, J. M. Alberdi *et al.*, *Macromolecules*, **32**, 4065 (1999).

Appendix AI

The Rouse model

AI.1 Formulation and normal modes

The best known and most widely employed model for polymer melt dynamics is the Rouse model [1,2]. In the Rouse model, the polymer chain is treated as a Gaussian random coil comprising a set of N beads, each connected to its immediately preceding and following neighbor by harmonic entropic springs with force constant k. Excluded volume and inertial effects are omitted and hydrodynamic interactions are treated with a simple frictional force on each bead determined by the segmental friction coefficient ζ. The beads also experience an external Gaussian white noise force. The equation of motion for the nth bead is given as

$$\zeta \frac{\partial \mathbf{r}_n(t)}{\partial t} = k[\mathbf{r}_{n+1}(t) - 2\mathbf{r}_n(t) + \mathbf{r}_{n-1}(t)] + \mathbf{g}_n(t), \qquad (AI.1)$$

where

$$\langle g_{n\alpha}(t) g_{m\beta}(t') \rangle = 2\zeta k_B T \delta_{nm} \delta_{\alpha\beta} \delta(t - t'). \qquad (AI.2)$$

The basic length scale is set by the statistical segment length, i.e., the root-mean-square distance between beads, σ. The solution of the equation of motion (eq. (AI.1)) for the Rouse chain with spring force constant $k = 3k_B T / \sigma^2$ yields [1]

$$\langle R^2 \rangle = N\sigma^2, \qquad (AI.3)$$

where $\langle R^2 \rangle$ is the mean-square end-to-end distance of the chain. The solution of the Rouse equation of motion is determined analytically by transformation to its eigenmodes, the Rouse modes, which are defined as [3]

$$\mathbf{X}_p(t) = \frac{1}{N} \sum_{n=1}^{N} \cos\left(p\pi \frac{n - 1/2}{N}\right) \mathbf{r}_n(t), \qquad (AI.4)$$

where p is the mode index ($1 \le p \le N$) and $\mathbf{r}_n(t)$ is the Cartesian coordinates of bead (segment) n. The self-correlation function for $\mathbf{X}_p(t)$ is given as

$$\langle \mathbf{X}_p(t) \cdot \mathbf{X}_p(0) \rangle = \frac{\langle R^2 \rangle}{8N(N-1)\sin^2\left(\frac{p\pi}{2N}\right)} \exp\left[\frac{-p^2 t}{\tau_R}\right]. \qquad (AI.5)$$

For large N and $p \ll N$

$$\langle \mathbf{X}_p(t) \cdot \mathbf{X}_p(0) \rangle = \frac{\langle R^2 \rangle}{2\pi^2 p^2} \exp\left[\frac{-p^2 t}{\tau_R}\right], \tag{AI.6}$$

where the Rouse time τ_R is given by

$$\tau_R = \frac{\zeta N^2 \sigma^2}{3\pi^2 k_B T}. \tag{AI.7}$$

where T is temperature. Hence the chain dimensions (N and σ) and segmental friction coefficient establish uniquely all time and length scales for the Rouse chain.

AI.2 Establishment of Rouse parameters for a real polymer

In order to apply the predictions of the Rouse model of the relaxational behavior to real polymer chains, it is necessary to parameterize the model for the polymer of interest. Note that a real polymer may be a physical polymer upon which experiments are performed, or a molecular mechanics model for a polymer upon which MD simulation studies are performed. Equating the mean-square end-to-end distance of the Rouse chain (eq. (AI.3)) with the measured (experiment or simulation) value of our real chain yields

$$N\sigma^2 = \langle R^2 \rangle_{\text{real}} = C N_{\text{bonds}} l^2, \tag{AI.8}$$

where C is the characteristic ratio of the real chain and l is the root-mean-square bond length. Real chains have subsegmental structure, and hence the number of statistical segments N is not equal to the number of backbone bonds N_{bonds}. We can equate the contour lengths of the chains:

$$N\sigma = N_{\text{bonds}} l. \tag{AI.9}$$

Combining eqs. (AI.8) and (AI.9) yields

$$N = N_{\text{bonds}}/C; \quad \sigma = Cl. \tag{AI.10}$$

A straightforward way to establish the monomer friction coefficient (and hence the Rouse time and all relaxation times) is through the center-of-mass diffusion coefficient D_{cm} of the polymer chain. The Rouse model yields [1]

$$D_{\text{cm}} = k_B T/N\zeta. \tag{AI.11}$$

Note that the Rouse model is an effective medium model that does not account for hydrodynamic interactions. As such it is most applicable to polymer melts where hydrodynamic interactions are largely screened. For polymer melts the segmental friction coefficient ζ represents the influence of the polymer matrix on the motion of a polymer segment.

AI.3 The viscoelastic response of a Rouse chain

In the Rouse chain forces act only between neighboring beads of the chain. In the continuum limit of the Rouse model the stress tensor can therefore be written as [1]

$$\sigma_{\alpha\beta}(t) = \frac{c}{N} \frac{3k_B T}{\sigma^2} \sum_{n=1}^{N} \left\langle \frac{\partial \mathbf{r}_{n\alpha}(t)}{\partial n} \frac{\partial \mathbf{r}_{n\beta}(t)}{\partial n} \right\rangle \tag{AI.12}$$

or equivalently in normalized coordinates

$$\sigma_{\alpha\beta}(t) = \frac{c}{N} \sum_{p=1}^{N} k_p \langle X_{p\alpha}(t) X_{p\beta}(t) \rangle, \tag{AI.13}$$

where c is the number density of segments. The time dependent shear relaxation modulus $G(t)$, given by

$$G(t) = \frac{V}{k_{\mathrm{B}}T} \langle \sigma_{\alpha\beta}(t) \sigma_{\alpha\beta}(0) \rangle, \tag{AI.14}$$

where V is volume, becomes

$$G(t) = \frac{V}{k_{\mathrm{B}}T} \frac{c^2}{N^2} \sum_{p=1}^{N} \left[k_{\mathrm{B}}T \exp\left(\frac{-p^2 t}{\tau_{\mathrm{R}}} \right) \right]^2 = \frac{c}{N} k_{\mathrm{B}}T \sum_{p=1}^{N} \exp\left(\frac{-2p^2 t}{\tau_{\mathrm{R}}} \right). \tag{AI.15}$$

Similarly, it can be shown that the recoverable creep compliance is given by [4]

$$J_r(t) = \frac{4}{\nu k_{\mathrm{B}}T} \sum_{p=1}^{\infty} \frac{1}{\theta_p^2} \left\{ 1 - \exp\left(-\frac{t}{\lambda_p} \right) \right\} \tag{AI.16}$$

for the continuous Rouse model. Here, ν is the number of Rouse segments per unit volume and

$$\lambda_p = \frac{\pi^2 \tau_{\mathrm{R}}}{\theta_p^2}. \tag{AI.17}$$

The numerical coefficients θ_p are determined from $\tan \theta_p = \theta_p$ with $p\pi < \theta_p < (p+1/2)\pi$ and $\theta_p \to (p+1/2)\pi$ for $p \to \infty$.

AI.4 Bead displacements and the coherent single-chain structure factor

Back transformation of the normal coordinates of the Rouse chain into real space yields the time-dependent Cartesian coordinates [1]

$$\mathbf{r}_n(t) = \mathbf{X}_0(t) + 2 \sum_{p=1}^{N} \cos\left(\frac{p\pi n}{N} \right) \mathbf{X}_p(t). \tag{AI.18}$$

Because the Rouse modes are orthogonal, the corresponding bead mean-square correlation functions are given as

$$\varphi_{mn}(t) = \langle (\mathbf{r}_n(t) - \mathbf{r}_m(0))^2 \rangle = 6 D_{\mathrm{cm}} t + |n - m| \sigma^2$$
$$+ \frac{4N\sigma^2}{\pi^2} \sum_{p=1}^{N} \frac{1}{p^2} \cos\left(\frac{p\pi n}{N} \right) \cos\left(\frac{p\pi m}{N} \right) \left[1 - \exp\left(\frac{-t}{\tau_p} \right) \right]. \tag{AI.19}$$

The coherent single-chain dynamic structure factor in the Gaussian approximation, which assumes that the bead displacements $\mathbf{r}_n(t) - \mathbf{r}_m(0)$ are Gaussian distributed, is given as

$$S_{\mathrm{coh}}(q, t) = \frac{1}{N} \sum_{m,n=1}^{N} \exp\left(-\frac{q^2}{6} \varphi_{mn}(t) \right), \tag{AI.20}$$

yielding

$$
s_{\text{coh}}(q, t) = \frac{1}{N} \exp\left(-q^2 D_{\text{cm}}t\right) \sum_{m,n=1}^{N} \exp\left(-\frac{q^2\sigma^2}{6}|n - m|\right.
$$
$$
\left. - \frac{2Nq^2\sigma^2}{3\pi^2} \cos\left(\frac{p\pi m}{N}\right) \cos\left(\frac{p\pi n}{N}\right) \left[1 - \exp\left(\frac{-t}{\tau_p}\right)\right]\right). \quad (\text{AI.21})
$$

References

[1] M. Doi and S. F. Edwards, *The Theory of Polymer Dynamics* (New York: Oxford University Press, 1988).
[2] P. E. Rouse, *J. Chem. Phys.*, **21**, 1273 (1953).
[3] W. Paul, G. D. Smith, and D. Y. Yoon, *Macromolecules*, **30**, 7772 (1997).
[4] H. Watanabe and T. Inoue, *J. Phys.: Condens. Matter*, **17**, R607 (2005).

Appendix AII

Site models for localized relaxation

In the polymeric glass, the general contour of a polymer chain is fixed but there are nevertheless numerous localized conformational motions characteristic of subglass relaxation processes. Due to the restrictions imposed by the fixed chain contours, motions such as dipolar reorientation are spatially restricted to the angular excursions attendant to the localized motion and the complete reorientational space associated with molecular tumbling is no longer available. This has an effect on the strength of the relaxation. In addition, energy differences between the conformational states involved can lead to strong Boltzmann-weighting temperature effects on relaxation strength [1,2]. For dipolar relaxation two simple cases are discussed here, the two-state and the three-state site models. In doing so the associated relaxation times are addressed. Mechanical relaxation is also taken up in terms of the two-state model.

AII.1 Dipolar relaxation in terms of site models

AII.1.1 Two-state model dynamics

A schematic energy diagram for a dipole that can reside in either of two energy states is shown in Figure AII.1. The net fluxes in an ensemble of such systems from state 1 containing N_1 dipoles to state 2 containing N_2 dipoles and vice versa are given by

$$dN_1/dt = -k_1 N_1 + k_2 N_2, \qquad (AII.1)$$

$$dN_2/dt = k_1 N_1 - k_2 N_2, \qquad (AII.2)$$

where k_1, k_2 are appropriate rate constants. The total number of dipoles, $N_1 + N_2 = N^0$ is conserved. These equations have solutions of the form $N \sim e^{-\lambda t}$ and therefore $dN/dt = -\lambda N$. This leads to the set of homogeneous equations,

$$(k_1 - \lambda)N_1 - k_2 N_2 = 0,$$
$$-k_1 N_1 - (k_2 - \lambda)N_2 = 0, \qquad (AII.3)$$

which in order to have non-trivial solutions requires that the secular determinant

$$\begin{vmatrix} (k_1 - \lambda) & -k_2 \\ -k_1 & (k_2 - \lambda) \end{vmatrix} = 0 \qquad (AII.4)$$

vanish, resulting in the secular equation

$$(k_1 - \lambda)(k_2 - \lambda) - k_1 k_2 = 0, \qquad (AII.5)$$

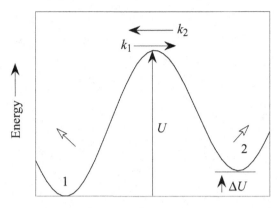

Figure AII.1 Energy diagram for a two-state dipole. The barrier from 1 to 2 is U and from 2 to 1 is $U - \Delta U$. The orientation of the dipole is different in the two states.

which has roots $\lambda' = 0$, $\lambda = k_1 + k_2$. Thus in terms of a *relaxation time* $\tau = 1/\lambda = 1/(k_1 + k_2)$, the equations

$$N_1 = N_1^e + \left(N_1^0 - N_1^e\right) e^{-t/\tau}, \tag{AII.6}$$

$$N_2 = N_2^e - \left(N_1^0 - N_1^e\right) e^{-t/\tau}, \tag{AII.7}$$

where N_1^e, N_2^e are the long-time equilibrium values of the populations and N_1^0 is the initial, $t = 0$, population in state 1, satisfy the differential equations and the initial conditions. N_1^0 must not be the same as N_1^e in order for relaxation to occur. The term $(N_1^0 - N_1^e)$ is a perturbation from equilibrium that causes relaxation to take place and is supplied by the application of the external electric field.

The equilibrium populations in terms of Figure AII.1 are given through Boltzmann weighting as

$$N_1^e/N^0 = 1/(1 + e^{-\Delta U/kT}), \tag{AII.8}$$

$$N_2^e/N^0 = e^{-\Delta U/kT}/(1 + e^{-\Delta U/kT}). \tag{AII.9}$$

Relaxation strength

Relaxation strength is conveniently represented by the dipolar correlation factor, g_P, in the Kirkwood–Onsager equation [3,4] (dipole number density $\overline{N}_P$, kT in cgs units, μ in electrostatic units so that dipole moments in Debye units (10^{-18} esu) can be directly used),

$$\varepsilon_R - \varepsilon_U = \left(\frac{3\varepsilon_R}{2\varepsilon_R + \varepsilon_U}\right)\left(\frac{\varepsilon_U + 2}{3}\right)^2 \frac{4\pi \overline{N}_P}{3kT} g_P \mu^2. \tag{AII.10}$$

In the present example it is assumed that the members of the ensemble making up the collection of two-state sites are randomly oriented with respect to each other and thus intermolecularly uncorrelated. In this circumstance, the correlation factor is given by [5]

$$g_P \mu^2 = \mu^2 - \langle \mu \rangle^2, \tag{AII.11}$$

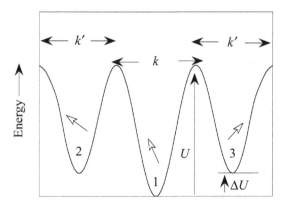

Figure AII.2 Schematic energy diagram for a three-state dipole. State 1 is more stable than the states 2 and 3, which are equally stable. The barrier and rate constant when going from 2 to 1 are the same as those when going from 2 to 3. The orientation of the dipole is different in the three states.

where $\langle \mu \rangle$ is the statistical mechanical average of the vector moment, μ. Thus

$$\langle \mu \rangle = p_1 \mu(1) + p_2 \mu(2), \tag{AII.12}$$

where $\mu(1)$, $\mu(2)$ are the dipole vectors in state 1 and 2 respectively and p_1, p_2 are the probabilities of finding the dipole vector in states 1 and 2. Those probabilities are given by eq. (AII.8) and eq. (AII.9). Writing out $\langle \mu \rangle^2$ explicitly using eq. (AII.12) gives

$$\langle \mu \rangle^2 = p_1^2 \mu^2 + 2 p_1 p_2 \mu(1) \cdot \mu(2) + p_2^2 \mu^2. \tag{AII.13}$$

Denoting $\mu(1) \cdot \mu(2)$ as $\mu^2 \cos \alpha$, where α is the angle between the dipole vector in its two states, invoking $p_1 + p_2 = 1$ in eq. (AII.13), and substituting the result into eq. (AII.11) results in

$$g_P = 2 p_1 p_2 (1 - \cos \alpha). \tag{AII.14}$$

Explicitly representing $p_1 p_2$ as Boltzmann factors through eq. (AII.8), eq. (AII.9) gives

$$g_P = 2(1 - \cos \alpha) e^{-\Delta U / kT} / (1 + e^{-\Delta U / kT})^2. \tag{AII.15}$$

It is noted that the correlation factor is restricted to values between 0 and 1. That is, the maximum value of the thermal population term is $1/4$, for $\Delta U = 0$. The maximum value of the geometric term is 4, for $\alpha = 180°$ reorientation.

AII.1.2 Three-state model

A schematic energy diagram for a three-state system where state 1 is more stable than states 2 and 3, which are equally stable, is shown in Figure AII.2.

The same procedure described above for the two-state system leads to three roots for the secular equation, the trivial solution $\lambda = 0$, and $\lambda_1 = 1/\tau_1 = 3k'$, $\lambda_2 = 1/\tau_2 = 2k + k'$. The relaxation strength through eq. (AII.11) with eq. (AII.12) generalized to

$$\langle \mu \rangle = p_1 \mu(1) + p_2 \mu(2) + p_3 \mu(3) \tag{AII.16}$$

is found to be

$$g_P = 2 p_1 p_2 (1 - \cos \alpha) + 2 p_2 p_3 (1 - \cos \beta) + 2 p_1 p_3 (1 - \cos \gamma), \tag{AII.17}$$

where α is the angle formed by the dipole in states 1 and 2, β the angle formed in states 2 and 3, γ the angle formed in states 1 and 3, and where the probabilities are given by

$$p_1 = 1/(1 + 2e^{-\Delta U/kT}), \tag{AII.18}$$

$$p_2 = p_3 = e^{-\Delta U/kT}/(1 + 2e^{-\Delta U/kT}). \tag{AII.19}$$

If states 2 and 3 are not of equal energy but have values ΔU_2, ΔU_3 respectively eq. (AII.17) is still valid, independently of the dynamics, but with the probabilities given by

$$p_1 = 1/(1 + e^{-\Delta U_2/kT} + e^{-\Delta U_3/kT}), \tag{AII.20}$$

$$p_2 = e^{-\Delta U_2/kT}/(1 + e^{-\Delta U_2/kT} + e^{-\Delta U_3/kT}), \tag{AII.21}$$

$$p_3 = e^{-\Delta U_3/kT}/(1 + e^{-\Delta U_2/kT} + e^{-\Delta U_3/kT}). \tag{AII.22}$$

AII.2 Mechanical relaxation in terms of site models

In the dipolar case the electric field perturbs the equilibrium distribution of dipole orientations and the approach to a new equilibrium is measurable via time domain or dynamic methods. In the mechanical analogy, the reorienting entities are not as precisely defined as a dipole, i.e., a bond possessing an electric moment. Conceptually, however, a bulk material can be envisioned as containing localized sites that consist of bond sequences that can undergo conformational transitions just as in the dipolar case. At the low temperatures characteristic of secondary, subglass relaxations the local chain segment that rearranges will not fit into its local environment of surrounding chains in the same way after the jump as before. Examination of the initial and final states of the examples in Figure 8.22 illustrates this point. This situation results in a strain field being created about the rearranging sequence or more realistically in a change in the strain field. Application of an external stress biases the equilibrium populations of the local conformational sequence through its coupling with the strain field. This allows observation of *anelastic relaxation* (Figure 1.2) [6,7].

The strain field decays as R^{-3} with distance [8] so that the product of the strain ϵ in a volume V, εV, is the conceptual basis of the relaxing entity. The strain field also has an angular dependence and both the strain and stress are actually vector quantities (see Appendix A1). In the interest of simplicity they are taken here as scalars. The product of the stress σ with εV, or $\sigma \varepsilon V$, can be considered to be the energy of interaction of the localized site with the stress. The total energy of a site n in the presence of the applied stress then can be regarded as

$$E_n = U_n - \sigma \varepsilon_n V,$$

where U_n is the site energy in the sense of Figures AII.1, AII.2. The equilibrium compliance, J_r, is

$$J_r = J_u + \overline{N} \langle \varepsilon \rangle_\sigma V/\sigma, \tag{AII.23}$$

where $\langle \varepsilon \rangle_\sigma$ is the statistical mechanical average of strains associated with each site in the presence of the stress σ, $\overline{N}$ is the number density of sites, and J_u is the unrelaxed compliance. The average, $\langle \varepsilon \rangle_\sigma$ is given by

$$\langle \varepsilon \rangle_\sigma = \sum_n \varepsilon_n e^{-(U_n - \sigma \varepsilon_n V)/kT} \bigg/ \sum_n e^{-(U_n - \sigma \varepsilon_n V)/kT}. \tag{AII.24}$$

The ordinary case of linear response of strain to stress is found from eq. (AII.24) through

$$\langle \varepsilon \rangle_\sigma = \left(\frac{\partial \langle \varepsilon \rangle}{\partial \sigma} \right)_{\sigma=0} \sigma, \tag{AII.25}$$

which gives

$$\langle \varepsilon \rangle_\sigma = (\langle \varepsilon^2 \rangle - \langle \varepsilon \rangle^2) \sigma V / kT, \tag{AII.26}$$

where $\langle \varepsilon^2 \rangle$, $\langle \varepsilon \rangle$ are the statistical mechanical averages in the absence of applied stress. The relaxation strength, using eq. (AII.23), is then expressed as

$$J_r - J_u = \overline{N} \langle \varepsilon \rangle V / \sigma = \overline{N} \left(\langle \varepsilon^2 \rangle - \langle \varepsilon \rangle^2 \right) V^2 / kT. \tag{AII.27}$$

This result is seen to be a close analogy to the dipolar situation as expressed in eq. (AII.11). The term $(\langle \varepsilon^2 \rangle - \langle \varepsilon \rangle^2)$ for two sites is written out as

$$(\langle \varepsilon^2 \rangle - \langle \varepsilon \rangle^2) = (p_1 \varepsilon_1{}^2 + p_2 \varepsilon_2{}^2) - (p_1 \varepsilon_1 + p_2 \varepsilon_2)^2, \tag{AII.28}$$

where the site probabilities are the Boltzmann weights

$$p_1 = 1/(1 + e^{-\Delta U / kT}), \tag{AII.29}$$
$$p_2 = e^{-\Delta U / kT} / (1 + e^{-\Delta U / kT}). \tag{AII.30}$$

Use of the $p_1 + p_2 = 1$ relation allows eq. (AII.28) to be reduced to

$$(\langle \varepsilon^2 \rangle - \langle \varepsilon \rangle^2) = p_1 p_2 (\varepsilon_1 - \varepsilon_2)^2$$

and hence eq. (AII.27) becomes, with the recognition that spatial averaging of a random uniaxial strain with a fixed uniaxial stress leads to a factor of $1/3$,

$$J_r - J_u = \overline{N} \langle \varepsilon \rangle V / \sigma = \overline{N} p_1 p_2 (\varepsilon_1 - \varepsilon_2)^2 V^2 / kT. \tag{AII.31}$$

The time dependence follows as in the two-state dielectric example above.

References

[1] G. Williams, *Adv. Polymer Sci.*, **33**, 159 (1979).
[2] G. D. Smith and R. H. Boyd, *Macromolecules*, **24**, 2731 (1991).
[3] N. G. McCrum, B. E. Read, and G. Williams, *Anelastic and Dielectric Effects in Polymeric Solids* (New York: Wiley, 1967; Dover, 1991), Chapter 3.
[4] H. Froehlich, *Theory of Dielectrics*, second edn (London: Oxford University Press, 1958).
[5] G. D. Smith and R. H. Boyd, *Macromolecules*, **24**, 2731 (1991), eq. (24) with μ for a single localized dipole instead of the molecular moment **m(x)**.
[6] C. Zener, *Elasticity and Anelasticity of Metals* (Chicago: University of Chicago, 1948).
[7] A. S. Nowick and R. S. Berry, *Anelastic Relaxation in Crystalline Solids* (New York: Academic, 1972).
[8] H. B. Huntington and R. A. Johnson, *Acta Metall.*, **10**, 281 (1962).